M000077533

ECOLOGIES OF HUMAN FLOURISHING

Religions of the World and Ecology Series

Other volumes in the Religions of the World and Ecology series published by Harvard Divinity School's Center for the Study of World Religions are available through Harvard University Press (www.hup.harvard.edu). The most recent of these is

- *Ecology and the Environment: Perspectives from the Humanities,* Donald K. Swearer, ed.

Earlier volumes for which Mary Evelyn Tucker and John Grimm were series editors; Lawrence E. Sullivan, general editor, CSWR publications; and Kathryn Dodgson, senior editor, CSWR publications, include:

- *Buddhism and Ecology: The Interconnection of Dharma and Deeds,* Mary Evelyn Tucker and Duncan Ryûken Williams, eds.
- *Christianity and Ecology: Seeking the Well-Being of Earth and Humans,* Dieter T. Hessel and Rosemary Radford Ruether, eds.
- *Confucianism and Ecology: The Interrelation of Heaven, Earth, and Humans,* Mary Evelyn Tucker and John Berthrong, eds.
- *Daoism and Ecology: Ways within a Cosmic Landscape,* N. J. Girardot, James Miller, and Liu Xiaogan, eds.
- *Hinduism and Ecology: The Intersection of Earth, Sky, and Water,* Christopher Key Chapple and Mary Evelyn Tucker, eds.
- *Indigenous Traditions and Ecology: The Interbeing of Cosmology and Community,* John A. Grim, ed.
- *Islam and Ecology: A Bestowed Trust,* Richard C. Foltz, Frederick M. Denny, and Azizan Baharuddin, eds.
- *Jainism and Ecology: Nonviolence in the Web of Life,* Christopher Key Chapple, ed.
- *Judaism and Ecology: Created World and Revealed Word,* Hava Tirosh-Samuelson, ed.

ECOLOGIES OF HUMAN FLOURISHING

Edited by Donald K. Swearer and Susan Lloyd McGarry

Center for the Study of World Religions
Harvard Divinity School
Cambridge, Massachusetts
Distributed by Harvard University Press
2011

Printed in the United States of America

Grateful acknowledgement is made for permission to use the following:

An earlier version of Donald K. Swearer's "Introduction: An Ecology of Human Flourishing" appeared as "An Ecology of Human Flourishing, 2008 Convocation Address," *Harvard Divinity Today* 4, no. 3 (Fall 2008): 9–12, Copyright ©2008 The President and Fellows of Harvard University. The present text is based on that version with permission.

Also in the "Introduction," an excerpt from "Please Call Me by My True Names," by Thich Nhat Hanh, *Being Peace*, 63–64, (Berkeley, CA: Parallax Press, 1997), is used by permission of Parallax Press, www.parallax.org.

Archana Venkatesan's "Love and Longing in the Time of Rain" contains excerpts of two translations by her, "*Tiruppāvai* 3" and Āṇṭāḷ in *Nācciyār Tirumoḻi* 8.1, from Archana Venkatesan, *The Secret Garland: Translations of Antal's Tiruppavai and Nacciyar Tirumoli* (New York: Oxford University Press, 2010); the excerpts are used by permission of Oxford University Press, Inc., www.oup.com.

The opening sections of Sallie McFague's "Cities, Climate Change, and Christianity: Religion and Sustainable Urbanism" draw heavily on excerpts from Chapter 7, "Where We Live: Urban Ecotheology," 121–139, in Sallie McFague, *A New Climate for Theology: God, the World, and Global Warming* (Minneapolis: Fortress Press: 2008) and are used by permission of Fortress Press.

Photos on cover used with permission.

From left to right, first row: Melbourne Australia 350 demonstration planting by Barbara Byrt; City view by Kimberley C. Patton; Man and wife in boat by Donald K. Swearer. Second row: Monks in prayer by Kristie Welsh; Iraqi children by Christiaan Briggs; Women in Leo market by Marco Schmidt. Third row: CSWR garden by Rebecca Kline Esterson; HDS student and staff by Justin Ide; Greek stream by Kimberley C. Patton

Cover Art and Design: Kristie Welsh

Library of Congress Cataloging-in-Publication Data

Ecologies of human flourishing / edited by Donald K. Swearer and Susan Lloyd McGarry.
p. cm. -- (Religions of the world and ecology series)
Includes bibliographical references and index.
ISBN 978-0-945454-45-8 (alk. paper)
1. Conduct of life. 2. Life. 3. Humanity. 4. Human beings. 5. Ethics.
I. Swearer, Donald K., 1934– II. McGarry, Susan Lloyd. III. Title. IV. Series.

BJ1548.E36 2011
201'.7--dc23

2011023814

Contents

Foreword

Our world is out of balance. Simply look: Atmospheric chemistry is tipping dangerously. The biosphere is out of sync. Eighty percent of the world's population lives on less than $10 per day, and more than 80 percent of us live in countries in which income disparities are increasing.[1] With little doubt, owing to the habitually profligate ways of those who can afford profligacy, human life on this planet appears to be running out of time.

On September 15, 2008, at the annual convocation of Harvard Divinity School, Donald Swearer courageously sounded not so much a note of alarm as one mostly of encouragement, positing that an "ecology of human flourishing" exists as an alternative to runaway pollution, consumption, economic inequality, and social decay.[2] He called for a new paradigm of moderation or "enoughness" to bring back into balance this careening Earth.

Over the following year, through the generous offices of Harvard Divinity School's Center for the Study of World Religions, which he then directed, Swearer began to marshal resources—both human (prodigious thinkers) and financial (modest sums)—to bring diverse analyses to this bright idea. In a series of lectures held at Harvard over the course of the 2009–10 academic year, scholars and students and laypeople of Cambridge and environs gathered monthly to listen and think, to weigh and converse, to struggle with the concept of how much longer and how well we can live together within this globe's finitude. I was a privileged witness to nearly all of these encounters.

Lectures and seminars at Harvard—I imagine this to be the case at hundreds of universities around the globe—are curiously ritualistic affairs. There is a room, a host, an introduction, a guest speaker, a speech, a respectful audience, questions and answers, a thank-you, good-byes, and a willy-nilly dispersal of newly formed seeds of ideas to settle hopefully on fertile ground. These gatherings of what Swearer now made plural as the "Ecologies of Human Flourishing" were similar but also somehow different.

The air in the room each time was heavily suggestive of people grappling with the intractable. These were not comfortable topics. No one could easily dismiss the consequences of the discussion, not only intellectually but also in the small, practical, subsequent decisions we would be making afterwards. Wouldn't eating one sandwich be "enough" at the reception following the talk? Surely, taking public transportation was the only ethical route home. In so deeply off-kilter a world, can our own simple and incremental lifestyle decisions really add up to something meaningful? When shall we begin to deny ourselves that trip to Europe? Or those new athletic shoes?

Swearer brought together as speakers: theologians, physicians, environmentalists, social scientists, and humanists. The mixing of scholarly perspectives is a hallmark of the endeavors of research centers at Harvard. But this mix was particularly notable because already-interdisciplinary centers on religion, on international affairs, and on the environment were working together for a meta-sharing of interdisciplinary perspectives. Traversing borders as though into previously unvisited or hardly visited countries, intellectual sojourners crisscrossed, coming upon new vocabularies and manners of thinking at each meeting. And many uncomfortable truths.

What follows, then, is a series of profoundly honest interrogations into the long-term sustainability of human life. Can an ethos of interdependence ever trump the drive toward individualism? Can an ethic of limit provide its own sort of fulfillment? Can incremental change make sufficient difference in a world already coming apart at the seams?

There are, indeed, a number of assumptions behind these questions. I think all of the speakers, and apparently much of the audience, agreed that we are living in the midst of truly serious ecological and social crises. Then there was the well-known caveat of Tancredi, to which most everyone in every audience would have assented: "If we want things to stay as they are, things will have to change."[3] Finally—and this is no stretch, at least not within the confines of Harvard and of Cambridge—we came to understand that a redefinition of what is enough (without causing undue pain, and yet necessitating a judicious redistribution of resources globally) is certainly a question of life and death.

So, what is to be done? Surely there are those many who aver that economic growth and technological development are the only ways forward—and the only way out of our current global predicaments. I have heard it said, for instance (and not unconvincingly), that low-input agriculture, farmers' markets, and local food that stays as far away as possible from modern commercial transport is just what most poor African rural dwellers have . . . and want to get beyond and leave behind as fast as they can. That is, we must grow and innovate out of our predicament. Does danger lurk, in other words, in the romance of ancient Indian pastoral civilizations, the Concord of Thoreau, and the exceptional lives of truly exceptional women and men of history whose examples are hardly replicable? Or, to get to the root of it, do we really believe that great expressions of human cooperation and altruism are scalable?

At the end of the year we left these halls of inquiry worried, analytically better suited to confront the answers, and, I am sure, more embracing of what it means to share and to lead—and, yes, to sacrifice.

For me, these seminars, whose wisdom I invite you to take in quietly and earnestly, spurred my resolve to think more and act further toward balance. Being among those fortunate to have more than enough, what will I do to bring a greater equilibrium to the world? Being among those who have enjoyed intellectual privilege, how do I share more fully that wealth that among all resources is peculiarly and hearteningly infinite: learning? How do we, as a species, redistribute and give up in order to flourish—and not to flourish merely personally but also communally, far beyond our own kind?

It is, truly, in our thinking holistically about the aggregate—the atmosphere, the biosphere, the human sphere—that we can contain multitudes and open our minds and hearts to finding a sustainable, flourishing balance.

Steven B. Bloomfield
Executive Director
Weatherhead Center for International Affairs
Harvard University
September 30, 2010

Notes

1. For the first figure, see Shaohua Chen and Martin Ravallion, "The Developing World Is Poorer than We Thought, but No Less Successful in the Fight against Poverty," Policy Research Working Paper 4703 (Washington, DC: World Bank, August 2009 revised version), 51; for the second, see United Nations Development Programme, *2007/08 Human Development Report* (New York: UNDP, November 2007), 25. Global Issues has gathered these and similar statistics together on its website, http://www.globalissues.org/article/26/poverty-facts-and-stats.
2. Donald Swearer coined this phrase. See his article "Buddhism and Ecology: Challenge and Promise," *Earth Ethics* 10, no. 1 (Fall 1998), for its first use.
3. Prince Tancredi is a character in the Italian novel *The Leopard*. See Giuseppe Tomasi di Lampedusa, *The Leopard*, trans. Archibald Colquhoun (New York: Pantheon Books, 2007), 28.

Preface

Donald K. Swearer

The phrase, "ecologies of human flourishing," was inspired by my association with the Thai monk Buddhadasa Bhikkhu (1904–1993), the founder of the monastery known as the Garden of Empowering Liberation *(Wat Suan Mokkhabalarama).* I met Buddhadasa for the first time in 1968, although I was introduced to his writings a decade earlier. Buddhadasa's influence on contemporary Thai Buddhism through his lectures, writings, and personal example cannot be overestimated, especially among Thai intellectuals and socially engaged Buddhists. His extensive body of published work includes not only brilliant and sometimes controversial interpretations of such seminal Buddhist concepts as emptiness, not-self, and dependent co-arising, but also addresses a wide range of social, political, economic, and environmental issues.

Buddhadasa inspired me and many others to contemplate questions of personal value, meaning, and identity; of what it means to be human and of human flourishing, not in splendid isolation, but in broad, inclusive terms. The Garden of Empowering Liberation, located outside of Chiaya in southern Thailand, exemplified an expansive sense and vision of what it means to be fully human. Buddhadasa embodied that vision for me. He often spoke about nirvana, Buddhism's summum bonum, not as something inaccessible, but as a way of being filled with understanding, equanimity, and a compassion that included all forms of life—an ecology of human flourishing.[1] His lecture attire was often mismatched cast-off monastic robes; cats and dogs and an occasional chicken wandered at his feet;

his provocative, learned remarks were replete with down-to-earth examples and engaging metaphors.

Buddhadasa was simultaneously exceptional and commonplace, just as his teachings ranged from sublime aspirations of the human spirit to basic human needs and activities. So it is with an ecology of human flourishing. The essays in this volume reflect such scope: inspirational, aspirational, pragmatic. The authors represent different fields and academic disciplines, hence the plural of the title, "ecologies of human flourishing." The very diversity of their contributions reflect not only the authors' backgrounds and interests, but also the variegated complexity of human flourishing when conceived in expansive ecological terms. The assemblage of papers in this volume includes a Malaysian political scientist and human rights activist (Chandra Muzaffar), a Christian constructive theologian (Sallie McFague), a medical anthropologist (Arthur Kleinman), an American literature Thoreauvian (Lawrence Buell), a historian of religions who specializes in premodern India (Anne Monius), one of today's foremost environmentalist writer-activists (Bill McKibben), and an equally well-known social psychologist and culture critic (Barry Schwartz). All of the contributors except Barry Schwartz participated in the 2009–10 CSWR Ecologies of Human Flourishing lecture series. Schwartz, a former colleague at Swarthmore College, whose books (*The Paradox of Choice*, *Costs of Living*) have influenced my understanding of "enoughness," graciously agreed to contribute an essay to this volume.

In my 2008 Harvard Divinity School convocation address, reprinted as the introduction to this book, I emphasized the urgency of transforming worldviews from what Sallie McFague characterizes as an "individualistic anthropology" to an "ecological anthropology." The essays in this volume reinforce this theme. For Chandra Muzaffar the economic crisis requires enactment of limits on consumption and recognition of the necessity of restraint and moderation. Bill McKibben calls for a "new logic," a set of goals and aspirations radically different from the constant emphasis on endless growth and economic expansion. In calling for a shift from an "individual-in-the-machine" model to an organic model for twenty-first-century urban living, McFague quotes Thomas Friedman, "What if the deep recession is telling us that the whole growth model we created over the past 150 years is simply unsustainable economically and ecologically?"[2] Anne Monius's essay on the urban culture of the Pallava Dynasty (seventh to ninth centuries) demonstrates that the Pallavas were deeply concerned about the relationship between the human, built environment—"second nature" in McFague's words—and "first nature," in particular regarding water resources on which human flourishing physically depends. The Pallava worldview linked the human and natural orders in ways that offer a lesson for citizens of the twenty-first century: individual virtue—in the Pallava case, of leaders (especially kings) responsible for water management—and the creation and sustenance of an ecology of human flourishing are organically interdependent. In our quest for sufficiency and sustainability we undervalue and often overlook the moral dimension of personal choice and lifestyle today.

An ecology of human flourishing embodies a worldview and a way of being and acting. For Thoreau, as Buell points out in his essay, living simply optimizes the possibility of human flourishing. In today's consumerist culture living simply too often smacks of moralistic naysaying. For Thoreau, however, it meant being able to live deliberately; slowing down the tempo of life to the point that, as Buell puts it, the "mind and the senses become attuned to squeeze the juice out of each moment." Buell asks of Thoreau, as in this volume David Lamberth asks of Sallie McFague and Ronald Thiemann of Chandra Muzaffar, whether the moral critique and spiritual aspiration of an ecology of human flourishing can really make a practical difference in the world beyond an individual lifestyle choice. In the case of Thoreau, Buell concludes that voluntary simplicity is a necessary precondition for human integrity and human flourishing, "even if not sufficient to preserve the world from ruin in the long run," and he finds in Albert Schweitzer an example of altruism that complements Thoreau's voluntary simplicity.

In global health initiatives such as the American Red Cross under Clara Barton, the World Health Organization under Halfdan Mahler as its director-general, UNICEF led by James Grant, and Partners In Health cofounded by Paul Farmer, Arthur Kleinman finds a partial answer to Buell's, Thiemann's, and Lamberth's interrogation of the practical value of an ecology of human flourishing. In Kleinman's words, "There is something in our deep and divided subjectivity that is the emotional and physiological basis of religious aspiration and commitment. Out of it comes remarkable, life-sustaining powers of caregiving that are foundational to the ethical passion required by many in global health, humanitarian intervention, and those fostering other prosocial forms of human flourishing."

Echoing Steve Bloomfield's foreword, it is my hope that the essays in this volume will provoke readers to think more deeply and critically about balance and living deliberately; about how we, as a species, can live within the limits of sufficiency and redistribute justly in order to enable the greater flourishing of all.

Notes

1. See Donald Swearer, "Buddhism and Ecology: Challenge and Promise," *Earth Ethics* 10, no. 1 (Fall 1998), for its first use in print.
2. Thomas Friedman, "The Inflection Is Near?" *New York Times*, March 8, 2009, WK12, as quoted in the opening paragraph of Sallie McFague's essay in this book.

Acknowledgements

Both editors would like to acknowledge the support of the Center for the Study of World Religions (CSWR) at Harvard Divinity School (HDS) in bringing this book to fruition. We would like to thank the new leadership of the CSWR, director Francis X. Clooney, S.J., and managing director Susan Abraham, for continued funding for the project, even after the editors left their positions at the Center. We would also like to express our deep gratitude to our former colleagues (Charles Anderson, Alicia Belair, Joe Cook, and Rebecca Esterson) for all their help in the work of the lecture series that provided the seed for this book and the process that brought it through its final stages. We would also like to thank Nancy Swearer for early editorial assistance (as well as ongoing support), and HDS MTS students Ann Marie Micikas and Ailya Vajid for early proofing help.

We would like to honor each of the individual contributors for their intellectual work, their commitment to these important ideas, and their patience with us during the editing process. We could only invite these contributors because of help from the cosponsors of the lecture series: the Harvard University Center for the Environment and its director, Daniel P. Schrag; the Weatherhead Center for International Affairs and its executive director, Steven B. Bloomfield; and the Initiative on Religion in International Affairs at the Kennedy School and its director, Monica B. Toft. This volume has been aided by the discussion after each lecture, both at the lecture site and informally over dinner following, which often brought to light new perspectives, so our gratitude also extends to the participants in these events.

Susan Lloyd McGarry would also like to thank Donald K. Swearer for his gentle leadership of the CSWR for six years and for his life as an exemplar of those virtues to which he calls us under the rubric of ecologies of human flourishing. As well, she would like to express her appreciation to Kathryn Dodgson, director of the Office of Communications at HDS, for important help in reframing the process of the book; and to her friends for their patience at her constant unavailability over the last few months.

Introduction: An Ecology of Human Flourishing

Donald K. Swearer

In 1975 four of my colleagues at Swarthmore College, where I was teaching at the time—a classicist, an anthropologist, a physicist, and a biologist—and I dreamed up the idea of teaching a foolishly ambitious course that we called "Patterns of Explanation."[1] The idea for the course came from our reading of Thomas Kuhn's provocative 1962 book, *The Structure of Scientific Revolutions,* leavened by Anthony F. C. Wallace's work on revitalization movements and Karl Jaspers's notion of *Achsenzeit*, or Axial Age.[2]

While a graduate student in physics at Harvard, Kuhn became enamored with the history of science and was lured away from theoretical physics into a distinguished academic career in the philosophy and history of science at Harvard, Berkeley, Princeton, and MIT. Central to Kuhn's project was, first of all, the concept of "paradigm," which he defined as "universally recognized scientific achievements that for a time provide model problems and solutions to a community of practitioners."[3] In particular, Kuhn addressed the nature of "paradigm shift," or revolutions, in which an older epistemological paradigm is replaced in whole or in part by a new one.

New paradigms emerge when the old ones encounter an accumulation of anomalies that a widely accepted paradigm can no longer explain, thereby creating an epistemological crisis. Kuhn included in his examples the transition from a Ptolemaic cosmology to a Copernican one; from the worldview of Newtonian physics to Einsteinian relativity; and from Lamarckism to Darwin's theory of natural selection.

Although Kuhn's book was about paradigm shifts in science, he found an analogy in political revolutions in which a crisis attenuates the role of political institutions: "In increasing numbers," contends Kuhn, "individuals become increasingly estranged from political life and behave more and more eccentrically within it. Then, as the crisis deepens, many of these individuals commit themselves to some concrete proposal [i.e., a new paradigm] for the reconstruction of society in a new institutional framework."[4]

Echoing Kuhn's claim, the sociologist Robert Bellah opened his keynote address at the August 2008 conference, "The Axial Age and Its Consequences for Subsequent History and the Present," with these remarks: "What has become clear to me in recent years is that the old dream [read: paradigm] of progress, which used to be assumed, is being replaced in popular culture by visions of disaster, [and] ecological catastrophe. . . . Never before have calls for criticism of and alternatives to the existing order seemed so urgent."[5]

Kuhn's paradigm model has little traction in today's poststructuralist, postmodernist academic ethos; however, I want to borrow from Kuhnian terminology to argue for a paradigm shift for the twenty-first century that I am calling an "ecology of human flourishing."

"Ecology of human flourishing" is an odd phrase, for it juxtaposes the term "ecology," whose main provenance is biological science, with "human flourishing," which derives from the terminology of virtue ethics. "Ecology," however, is often used rather loosely, not in the specific biological terms of the relations between living organisms and their environment, but referring in a much broader sense to the interdependent, interrelational nature of all things including both the natural and human worlds.

The causal import of such a worldview is reflected in the 1975 National Academy of Sciences Report: our world is a whole "in which any action influencing a single part of the system can be expected to have an effect on all other parts of the system."[6] I am using the term "ecology" in this broad sense and, when coupled with "human flourishing," the phrase incorporates both a worldview and a lifestyle. An ecology of human flourishing, then, is *an understanding of the world as organically interrelated and interdependent* and is *a way of being and acting in the world informed and motivated by such a worldview.*

E. O. Wilson writes specifically about human dependence on nature, not from a utilitarian point of view, but from the perspective of what it means to be fully human. Apprehensive of the danger that computer-based information technology might foster the belief "that the cocoons of urban and suburban material life are sufficient for human fulfillment," Wilson contends, in his book *The Creation: An Appeal to Save Life on Earth*, that "Human nature is deeper and broader than the artifactual contrivance of any existing culture. The spiritual roots of *Homo sapiens* extend deep into the *natural world*. . . . We will not reach our full potential without understanding the origin and hence meaning of the aesthetic and religious qualities that make us ineffably human."[7]

Inspirations for the Ecology of Human Flourishing

> The vastness of the universe which you contemplate in a star-lit night becomes even vaster when you look at yourself as part of this universe and when you begin to realize that it is you who are this universe, contemplating itself.
>
> —Ernesto Cardenal[8]

As noted in the Preface, the expression "ecology of human flourishing" was initially inspired by my friendship with the Thai monk Buddhadasa Bhikkhu—a major figure in contemporary Thai Buddhism—and by his writings on the environment. In poetic prose Buddhadasa writes: "The entire cosmos is a cooperative. The sun, the moon, and the stars live together as a cooperative. The same is true for humans and animals, trees and the earth. . . . When we realize that the world is a mutual, interdependent, cooperative enterprise . . . then we can build a . . . noble environment. If our lives are not based on this truth, then we shall perish."[9]

Like E. O. Wilson, for whom the spiritual roots of *Homo sapiens* lie in the natural world, Buddhadasa valued the Garden of Empowering Liberation as a natural setting and place where *Homo sapiens* might—in Wilson's language—become more "ineffably human." "The deep sense of calm," writes Buddhadasa, "that nature provides through separation from the stress that plagues us in the day-to-day world protects our heart and mind. The lessons nature teaches us lead to a new birth beyond the suffering caused by our acquisitive self-preoccupation."[10]

Buddhadasa intended the Garden of Empowering Liberation not as a retreat from the world, but as a place where all forms of life—humans, animals, and plants—might live as a cooperative microcosm of the larger ecosystem, and as a community where humans might develop an ecological ethic based on the values of mindfulness, moderation, simplicity, and nonacquisitiveness. Buddhadasa firmly believed that technology and government alone cannot solve the eco-crisis, but that it requires a transformation of values and of lifestyle.

My work on Buddhadasa's environmental writings was occasioned by a conference on Buddhism and ecology organized at the Center for the Study of World Religions in 1996. It was the first of several conferences on religion and ecology held at the CSWR that led to the founding of the Forum on Religion and Ecology and to the emerging field of religion and ecology studies. The papers published from these conferences encompass a variety of topics. Some are critical of the religions' historical record regarding the environment, but the majority interrogate religious traditions as a possible resource for the development of an environmental ethic. Mary Evelyn Tucker, the cofounder of the forum, optimistically sees the world's religions entering an "ecological phase," a recentering of the human within the myriad species with whom we share the planet, in recognition that human flourishing is inextricably linked to the flourishing of the entire biotic community, and the extension of care and compassion toward what she terms "the great fecundity of life."[11]

There is, of course, nothing unique about holistic worldviews that stress the interconnectedness of all life-forms, whether grounded in a transcendent reality or not. In her book *The Great Transformation: The Beginning of Our Religious Traditions*, Karen Armstrong, following Karl Jaspers, notes that the "Axial sages," who propounded the great world religions and philosophies between 800 and 220 BCE, advocated an ethic of compassion based on such a paradigm. Her normative concern, however, aims more at the present than the past. She concludes her book with these words: "Today we are living in a tragic world where, as the Greeks knew, there can be no simple answers; the genre of tragedy demands that we learn to see things from other people's point of view. If religion is to bring light to our broken world, we need, as Mencius suggested, to go in search of the lost heart, the spirit of compassion that lies at the core of all of our traditions."[12]

Ecological Anthropology

> What if we begin to realize that the community model—the model in which human individuals must fit into a just, sustainable planet—is a necessity? What if we wake up from our dream of individualistic glory . . . and realize that either we will all make it together, or none of us will make it?
>
> —Sallie McFague[13]

An ecology of human flourishing joins company with theologian Sallie McFague's "ecological anthropology." For McFague, ecological anthropology represents a paradigm shift from the "individualistic anthropology" that she sees as the dominant model of Western democratic, capitalist societies. She argues that most people in Western capitalistic democracies think of themselves first as individuals rather than as members of a community, especially a natural or planetary community. This individualistic outlook promotes a deeply ingrained competitive mentality that tends to see others, "both humans and other life-forms, as resources toward one's goal of self-sufficiency."[14]

For McFague, the crisis of global warming is linked to this "canopy" of individualism. The creation of a just and sustainable planet, she argues, calls for a revolutionary challenge to the paradigm of individualism that has influenced the unconscious and semiconscious assumptions of three pillars of American society—religion, economics, and government.[15]

McFague identifies the new paradigm as "communitarian": a model that emphasizes "our interrelationship and interdependence with all other human beings and other life-forms."[16] She calls Christians to an "ecological catholicity" and to "cruciform living" in "solidarity with those billon or so human beings who exist on a dollar a day," animals facing the loss of habitat, and "a deteriorating planet" "dying from excessive energy use."[17] Her paradigm calls for a shift from the neoclassical economic model that rules the global marketplace to an ecological or planetary model based on the assumption that we "need one another to survive and flourish."[18]

Communality

> If you look . . . deeply, with the eyes of those who are awake you see not only the cloud and the sunshine in it [a piece of paper] but that everything is here: the wheat that became bread for the logger to eat—the logger's father—everything is in this sheet

> of paper. . . . The presence of this tiny sheet of paper proves the presence of the whole cosmos.
>
> —Thich Nhat Hanh[19]

Like McFague's paradigm, an ecology of human flourishing highlights "commun-ality" and a lifestyle of "enough-ness." *Communality* in my paradigm has a planetary reach that includes, but is not limited to, human community. "Community" is a buzz word in many places, including Harvard Divinity School, where I first wrote these words. There we talk a lot about the need for community, how to create community, community spaces, and so on. Robert Putnam, Professor of Public Policy at Harvard and former Dean of Harvard Kennedy School, who has focused his research on civic community, has made the phrase "bowling alone" a symbol of the loss of community or "social capital" in contemporary America.[20] Perhaps it is this sense of the loss of community that motivates us to strive to recreate it.

For Bishop Desmond Tutu, community defines our very humanity: "We are made for fellowship," observed Tutu, "because only in a vulnerable set of relationships are we able to recognize that our humanity is bound up with the humanity of others" (*Sermon in Birmingham Cathedral*). Bishop Tutu's communitarian theology, based on the African notion of *Ubuntu* (humanity), inspired the title of the 2008 documentary film *I Am Because We Are.*[21] The film is about the millions of orphans in the African country of Malawi who have lost parents and siblings to HIV and AIDS. Human flourishing in my paradigm extends this sense of "I-Am-Because-We-Are" to the community of nature as well.

Enoughness

> Peace and survival of life on earth as we know it are threatened by human activities which lack a commitment to humanitarian values. . . .
>
> Future generations . . . will inherit a vastly degraded planet if world peace does not become a reality and destruction of the natural environment continues at the present rate.
>
> —His Holiness the Dalai Lama[22]

An ecology of human flourishing also calls for a lifestyle of moderation or *enoughness*. There is nothing new about the value of enoughness. Thrift, moderation, and avoiding excess are classic virtues. The Buddhist tradition identifies itself as a Middle Way between the extremes of excessive indulgence and hair-shirt self-denial. And today, even op-ed columnist David Brooks, in his June 10, 2008, *New York Times* column, attacks what he refers to as our excessive "financial decadence."[23] The philosophy of enoughness, however, speaks to more than personal lifestyle or how we use or do not use money.

Several of environmentalist Bill McKibben's systemic issue books can be framed within the context of the principle of enoughness: *The End of Nature* (1989) is about the excessive release of carbon dioxide into the atmosphere that leads to global warming; *The Age of Missing Information* (1992) is about the flood of electronic information, especially through television, that endangers a deeper, contemplative understanding of who we are and the nature of the world we inhabit; *Enough: Staying Human in an Engineered Age* (2003) queries the limits and potential excesses of genetic engineering, robotics, and nanotechnology for what it means to be human; and *Deep Economy: The Wealth of Communities and the Durable Future* (2007) challenges the prevailing paradigm that "more is better."

In his book *The Paradox of Choice*, the social psychologist Barry Schwartz, critiques the excesses of the consumer paradigm that produces an overwhelmingly time-consuming and stultifying excess of choice, to the detriment of human flourishing. He begins the book:

> Scanning the shelves of my local supermarket recently, I found 85 different varieties and brands of crackers. As I read the packages, I discovered that some brands had sodium, others didn't. Some were fat-free, others weren't. They came in big boxes and small ones. They came in normal size and bite size. . . . [N]ext to the crackers were 285 varieties of cookies. Among chocolate chip cookies, there were 21 local options. Among Goldfish . . . there were 20 different varieties to choose from.

And goes on:

> I left the supermarket and stepped into my local consumer electronics store. Here I discovered:
>
> 45 different car stereo systems, with 50 different speaker sets . . .
> 42 different computers . . .
> 27 different printers . . .
> 110 different televisions . . . [And so on and so on].[24]

Schwartz's point is obvious—that the excesses of choice can be not only inordinately time-consuming, but mind-numbing and immobilizing. In other words, excessive choice undermines rather than promotes human flourishing.

The principle of enoughness embraces not only our personal lifestyles—from whether and what we watch on TV to our shopping habits—but also the ways we relate to and interact with others and with the natural environment. Maintaining a sustainable planet is a principle of enoughness. "Living simply that others may simply live" is a practical expression of the principle of distributive eco-justice, but is equally an expression of the virtues of care and compassion that flow from *Ubuntu,* "I am because all of us are." These virtues, as Mencius recognized, are constitutive of what it means to be human. Similarly, for Arthur Kleinman, who holds professorships at Harvard in anthropology and at the medical school and directs the Asia Center, caregiving is essential to human flourishing. It is an expression of our ethical and religious aspiration to remake the world in the midst of the contradictions, injustices, and the problematic of our daily lives.[25] It was one of the great felicities of my time as CSWR director that so many of those who have informed my thinking about an ecology of human flourishing were willing to contribute further to it, by participating in the lecture series and in this book.

Connecting the Ethic and the Worldview

For me, Thich Nhat Hanh's poem "Please Call Me by My True Names" provides a particularly moving example of the connection between the worldview and the ethic of an ecology of human flourishing. It is a poem that he wrote after receiving a letter about a twelve-year-old girl on a small refugee boat fleeing Vietnam after the war, who was raped by a sea pirate and who then leapt into the ocean and drowned herself. The poem moves from the interconnectedness Nhat Hanh feels with nature—

> Look deeply: I arrive in every second
> to be a bud on a spring branch,
> to be a tiny bird, with wings still fragile,
> learning to sing in my new nest,
> to be a caterpillar in the heart of a flower

—to his identification with both the girl and the pirate—

I am the 12-year-old girl, refugee
        on a small boat,
who throws herself into the ocean after
being raped by a sea pirate,
and I am the pirate, my heart not yet capable
        of seeing and loving.

—and concludes,

Please call me by my true names
so I can hear all my cries and laughter at once,
so I can see that my joy and pain are one.

Please call me by my true names,
        so I can wake up,
and so the door of my heart can be left open,
the door of compassion.[26]

Many of us share Robert Bellah's sensibility that alternatives to the existing order have never seemed so urgent; or, from a Kuhnian perspective, that we have arrived at a crisis moment in human history when an increasing number of us are prepared to commit ourselves to a new paradigm. The sentiments embodied in the writings of Thich Nhat Hanh, the Dalai Lama, Ernesto Cardenal, and Sallie McFague underscore the twin pillars of an ecology of human flourishing: the world as organically interrelated and interdependent; and an ethical imperative that proceeds from it.[27] The worldview and ethic of an *ecology of human flourishing* so movingly and compelling expressed by those writers are not new. But it would be new—revolutionarily new—if, in the future, an ecology of human flourishing became the operative system of understanding of and for this century.

Notes

1. This essay is based on Donald Swearer's convocation address, given on September 15, 2008, to the Harvard Divinity School community at the opening of the 2008–09 academic year. This version is slightly revised from that printed in *Harvard Divinity Today* 4, no. 3 (Fall 2008): 9–12, and is based on that version with permission.
2. Thomas S. Kuhn, *The Structure of Scientific Revolutions* (Chicago: University of Chicago Press, 1962); Karl Jaspers, *The Origin and Goal of History,* trans. Michael Bullock (New Haven: Yale University Press, 1953); and Anthony F. C. Wallace, "Revitalization Movements," *American Anthropologist* 58 (1956): 264–281.
3. Kuhn, *Structure of Scientific Revolutions,* viii.
4. Ibid., 93.
5. Robert Bellah, "Is Critique Secular? The Renouncers," keynote speech, August 2008 conference, "The Axial Age and Its Consequences for Subsequent History and the Present," Max Weber Center for Advanced Cultural and Social Studies in Erfurt, Germany, condensed version posted by Bellah on *Social Science Research Council Blogs. The Immanent Frame: Secularism, Religion, and the Public Square,* posted August 11, 2008, www.ssrc.org/blogs/immanent_frame/2008/08/11/the-renouncers/.
6. National Research Council/National Academy of Sciences, "Understanding Climate Change: A Program for Action," NAS/NRC Report (Washington, D.C.: National Academies Press, 1975).
7. E. O. Wilson, *The Creation: An Appeal to Save Life on Earth* (New York: W. W. Norton, 2006), 12.
8. Ernesto Cardenal, *To Live Is to Love* (New York: Herder and Herder, 1972), 149.
9. Buddhadasa Bhikkhu, *Phutasasanik Kap Kan Anurak Thamachat* (Buddhists and the Care of Nature) (Bangkok: Komol Thimthong Foundation, 1990), 35. Translation is mine.
10. Buddhadasa Bhikkhu, *Siang Takon Jak Thamachat* (Shouts from Nature) (Bangkok: Sublime Life Mission, 1971), 6–7. Translation is mine.
11. Mary Evelyn Tucker, *Worldly Wonder: Religions Enter Their Ecological Phase* (Chicago: Open Court, 2003).
12. Karen Armstrong, *The Great Transformation: The Beginning of Our Religious Traditions* (New York: Alfred A. Knopf, 2006), 476.
13. Sallie McFague, *A New Climate for Theology: God, the World, and Global Warming* (Minneapolis: Fortress Press, 2008), 44.
14. Ibid., for "ecological" and "individualistic" anthropology, see chapter 3, "Who Are We? Ecological Anthropology," 43–59; quote in last sentence is from 44.
15. Ibid., 46.
16. Ibid., 29.
17. Ibid., 35.
18. Ibid., 34.
19. Thich Nhat Hanh, *Being Peace* (Berkeley, CA: Parallax Press, 1987), 46–47.
20. Robert D. Putnam, *Bowling Alone* (New York: Simon and Schuster, 2000).

21. See Desmond Tutu, sermon in Birmingham Cathedral, April 21, 1988, transcript published by the Committee for Black Affairs, Diocese of Birmingham. For more on Tutu's theology of *Ubuntu*, see Michael Battle, *Ubuntu: I in You and You in Me,* with a foreword by Tutu and extensive quotes from Tutu's sermon (New York: Seabury Books, 2009). There is also a companion book to the movie: see Kristen Ashburn, *I Am Because We Are* (Brooklyn, NY: Powerhouse Cultural Entertainment, 2009).
22. Bstan-'dzin-rgya-mtsho, Dalai Lama XIV, "An Ethical Approach to Environmental Protection," in *Buddhist Perspectives on the Ecocrisis*, ed. Klas Sandell (Kandy, Sri Lanka: Buddhist Publication Society, 1987), 8.
23. David Brooks, "The Great Seduction," June 10, 2008, *New York Times,* op-ed column, online at http://www.nytimes.com/2008/06/10/opinion/10brooks.html.
24. Barry Schwartz, *The Paradox of Choice* (New York: Harper Collins, 2004), 9, 12–13.
25. Arthur Kleinman, "Today's Biomedicine and Caregiving: Are They Incompatible to the Point of Divorce?" (University of Leiden, November 26, 2007).
26. Thich Nhat Hanh, *Being Peace*, 63–64, used with permission from Parallax Press, www.parallax.org.
27. Longer selections from each were read as part of the Divinity School convocation ceremony.

Does Thoreau Have a Future? Reimagining Voluntary Simplicity for the Twenty-First Century

Lawrence Buell

The short answer to my title question is "yes, but. . . ."[1] Yes, Henry David Thoreau is most definitely still alive and well as prototype and poster child for voluntary simplicity advocates today, just as in his own time—and not merely in his home region or country, but even on the other side of the earth. A recent collection from Australia and New Zealand, *Voluntary Simplicity: The Poetic Alternative to Consumer Culture*, invokes Thoreau's example far more prominently than that of any other figure in world history from Buddha to Jesus to Gandhi.[2] Two very large "buts," however, are that Thoreau's continuing value as a model for ethical recalibration in the here and now must hinge on right understanding of what he does and does not have to offer by way of guidance and, quite apart from that, on the power of voluntary simplicity itself—Thoreauvian or otherwise—to serve as a bona fide game changer for individual lives and for human and planetary needs.

Before elaborating, conscience compels me to pause for a moment of true confessions. Voluntary simplicity is an ethos toward which I am both strongly drawn and acutely sheepish about passing judgment, as if pretending to speak from a higher plane of moral authority than the scribes and the pharisees. "What right have I to write on Prudence, whereof I have little?" So starts an essay by Ralph Waldo Emerson on a kindred theme.[3] I know the feeling. How fitfully have I practiced what I am about to preach, despite my sincere respect *in principle* for the version of it that I shall be defining here. I cannot deny living in a commodious home in an upscale exurb, despite the distinct remembrance of fervently agreeing in youth with my then-best friend that the curse of humankind was superfluous accumulation of property. I cannot deny belonging near the top of the 20 percent of earth's population that—according to one Worldwatch report—earns *on average* more than thirty times as much as the bottom 20 percent and two-thirds of total global income.[4] But at least I can take consolation in knowing I am not alone, that many if not most of *you* are in the same boat. Wendell Berry declares, "A protest meeting on the issue of environmental abuse is . . . a convocation of the guilty."[5] So too with a discussion of voluntary simplicity at Harvard University, the imperium of U.S. higher education. In a mid-1990s poll of American attitudes toward materialism, 83 percent of respondents affirmed that the United States consumes too much and still more agreed that "protecting the environment will require major changes in the way we live," as compared to only 28 percent who claimed that they themselves had, within the past five years, changed their lifestyle accordingly.[6] Quite a gap.

Up to a point, some such opinion-versus-behavior slippage holds true throughout American history, perhaps for all time. The prophets of the great world religions stood for a degree of moderation or self-restraint that institutionalized Confucianism, Christianity, and Buddhism, to name a few, have never attained, except fleetingly. In the secular arena, U.S. history makes an especially riveting exhibit insofar as American promise has for centuries been linked both to striking it rich and to the dream of a purified social order. As historian David Shi writes in his 1985 book on the subject, although the dream of the simple life—Puritan, Quaker, Shaker, Transcendentalist, and so forth—took hold in early colonial times and has remained forever after deeply embedded in American culture, it also has a way of "becoming enmeshed in its opposite," even as it has "served as the nation's conscience, reminding Americans of what the founders had hoped they would be and thereby providing a vivifying counterpoint to the excesses of materialist individualism." On that account, he predicts that "the simple life will persist both as an enduring myth and as an actual way of living," but offers no guarantee that any particular movement or initiative will endure for very long.[7] Shi's own lifeline makes a provocative cameo exhibit. Soon after completing *The Simple Life* in the pastoral semi-utopia of Davidson College, he assumed the chairmanship of his department, wrote another book on the transition during the American nineteenth century from idealism to realism, and ascended to the presidency of Furman University.[8]

But I do not want to sound as if I am espousing a uniform law of noble aspiration undermined by duplicitous slippage that holds true for all times and places. To see history that way is both unnecessarily depressing and not depressing enough. To take the bad news first, there is clearly something distinctively ominous about the trend line of the past century, especially since World War II: accelerating technosocial change increasingly imperils both planetary natural resources and the earth's human have-nots, and yet, ironically, this trend has been accompanied by the institutionalization of the conjoined assumptions that economic growth is essential to social well-being and that mass consumption is a key driver—if not *the* key driver—of that growth process. Harvard historian Lizbeth Cohen aptly terms the American variant of this idea "the Consumer Republic," born during the Depression years but launched full blast after World War II: a persuasion, as she puts it, that "promised the socially progressive end of economic equality" without establishing "politically progressive means of redistributing existing wealth," the argument being that "an ever growing economy built around the dynamics of increased productivity and mass purchasing power would expand the overall pie without reducing the size of any of the portions."[9] Though this grand vision of materialistic apotheosis faltered after the 1960s, when the Vietnam War and domestic strife derailed Lyndon Johnson's Great Society program, Cohen rightly points out that "patriotic shopping," as another commentator terms it, continues to be held up in the twenty-first century as a remedy for economic recession or the traumas of 9/11 and wars in the Middle East.[10]

Meanwhile, it is surely no coincidence that the benefits of economic prosperity themselves have lately become a subject of debate for economists and psychologists as well as ethicists. Does higher income really make individuals and societies happier? The answer seems to be broadly yes, but, predictably, that each increment matters proportionately more at the lower end of the scale and that "extra income increases happiness less and less as people get richer."[11] Even if Bill McKibben and others overstate the case in claiming that past the $10,000 per capita mark the correlation between income and happiness "disappears,"[12] it is striking that in a 2005 poll the cohort of "Forbes richest Americans" weighed in only very slightly higher in their rating of life satisfaction than "Traditional Masai" and "Pennsylvania Amish."[13] Moreover, the longitudinal data show the contemporary United States as a global outlier; for despite a modest rise in real household income for all economic brackets between 1972 and 2005, reported national happiness stayed flat, increasing slightly only for the top two quintiles, which received the biggest increments in income.[14] Particularly in light of today's gross economic inequality and overstrain on the earth's resources, the fact that the Masai and the Amish, whose ecological footprints are so much smaller than those of the Forbes millionaires, claim to feel so good—relatively speaking—while living on so much less makes a strong *prima facie* case for an ethics of voluntary simplicity as an antidote to the prodigalities of affluence.

Voluntary simplicity as a self-conscious persuasion under that particular name was in fact born concurrently with and as a counterweight to the incipient idea of the Consumer Republic, and has evolved roughly in tandem as a critique of the late twentieth century's mainstream gospel of wealth. Social philosopher Richard B. Gregg, an admirer of Gandhi best known in his day for his advocacy of nonviolence, first coined the term in a 1936 pamphlet written against "Henry Ford's idea that civilization progresses by the increase in the number of people's desires and their satisfaction." Gregg seeks to keep runaway consumption within bounds by propagating an ethic of "singleness of purpose, sincerity and honesty within, as well as avoidance of exterior clutter, of many possessions irrelevant to the chief purpose of life." Voluntary simplicity "means an ordering and guiding of our energy and our desires, a partial restraint in some directions in order to secure greater abundance of life in other directions." As such, it can also be "a mode of psychological hygiene," as he puts it, quaintly but tellingly. "Just as eating too much is harmful to the body," so, too, excessive attachment to material property can damage psychological health.[15] Gregg is quite aware that he risks sounding out of touch with the "mental climate" of the times, which holds simplicity to "be a foible of saints and occasional geniuses, but not something for the rest of us," and he compounds this risk by never once mentioning the then-crisis of the Great Depression. But he takes advantage of that moment by stressing several times the fragility of the sense of power and satisfaction conferred by "ownership of things."[16]

Although Gregg was the first to give the ethos which he described the name by which it is now best known, he knew perfectly well that he was not its first exponent. He invokes a host of precedents ancient and modern—from Moses, Buddha, and Confucius to Gandhi, as well as Lenin (!) and the Japanese peace activist Toyihiko Kagawa. Of Thoreau, Gregg makes no mention—the only American on his list is eighteenth-century Pennsylvania Quaker John Woolman. This selectivity may reflect not just Gregg's own religious affiliation but also a desire to immunize voluntary simplicity against the charges of bad citizenship, ascetic extremism, and sheer crankiness that still bedeviled Thoreau's reputation, even though he was on the verge of gaining his present stature as a major American writer and reform thinker. That two-part story of eclipse and rehabilitation makes a good next step for assessing Thoreau's standing today as a more prominent early spokesman for voluntary simplicity than Gregg, and beyond that, for assessing the intrinsic complexities, promise, *and* also vulnerabilities of voluntary simplicity ethics.

That Thoreau could matter much more to Gregg's hero Gandhi than to Gregg himself was partly Thoreau's own doing but was more due to the ironic legacy of Thoreau's mentor Emerson, whose distinctly tempered 1862 funeral address for the man then known chiefly as his disciple stood for the next half century as the definitive image of Thoreau in American memory. In late-nineteenth-century Britain, admiration for Thoreau as a progressive thinker in matters of politics, diet, and general lifestyle was nurtured by Fabian Socialists and other Victorian radicals with whom Gandhi came in contact, while the prevailing vision of Thoreau stateside, except among natural history buffs and anarchist fringe groups, was shaped by Emerson's image of the standoffish hermit-stoic whom one could admire only with a certain shudder, because "the severity of his ideal interfered to deprive him of a healthy sufficiency of human society."[17] I suspect that Emerson set out to *honor* Thoreau as a kind of latter-day Socrates, in a semi-reprise of his earlier essay on Plato, but that the cooling of their friendship in later years, as Emerson became more the man of affairs and Thoreau the idiosyncratic local character, drove Emerson to describing him as Diogenes instead.

What his portrait screens out is partly suggested by sociologist Robert Wuthnow's enlistment of Thoreau in a thoughtful book about underappreciated traditions of moral restraint embedded in the American version of the Protestant work ethic, *Poor Richard's Principle.* Wuthnow distinguishes two such strands, which he calls "ascetic moralism" and "expressive moralism," both of which he argues still persist today albeit in attenuated form. The first operates from a "fixed set of morally prescribed rules of behavior."[18] The second operates from the quest for modes of work fulfilling to the spirit. Wuthnow places Thoreau as a quintessential, iconic example of the second type, expressive moralism.[19] Emerson might have agreed in principle; he too commends Thoreau's refusal to enter "any narrow craft or profession" in order to follow "a much more comprehensive calling, the art of living well."[20] But Emerson spins this as Thoreau the obsessive naysayer and self-denier, whereas for Wuthnow, it is Thoreau as self-fulfiller.

However well Emerson knew Thoreau the person, Wuthnow's version is the better key to explaining the lure of *Walden*—then and now, Thoreau's most influential book. In *Walden*'s reminiscence of the author's two-year homesteading experiment, simplification of the terms of existence down to the bare essentials is offered less as austerity for its own sake than in order to optimize the possibility of human flourishing. The book seeks to convince us that "to maintain one's self on this earth is not a hardship but a pastime, if we will live simply and wisely."[21] Although Emerson likened Thoreau to a medieval monk, *Walden* claims that his business-as-usual workaholic townsmen are the ones who subject themselves to grotesque monkish penances, whereas he has managed to free himself for the maximum of "vacation from humbler toil" by sticking to necessities and dispensing with superfluities.[22] Even more fundamental than Thoreau's gospel of "simplify, simplify, simplify" is what he calls living "deliberately," that is, slowing down the tempo of life to the point that mind and senses become attuned to squeezing the juice out of each moment—from sunrise to sunset and beyond—so that even "the faint hum of a mosquito making its invisible and unimaginable tour through my apartment at earliest dawn when I was sitting with doors and windows open" finds its place in a ritual of awakening experienced as intensely meaningful.[23] The countless touches of this same kind throughout make *Walden* a testament to the pleasures of what today's voluntary-simplicity-speak calls "downshifting."[24]

One of the first readers to perceive this side of Thoreau was the novelist Virginia Woolf, in her charming tribute for the centennial of Thoreau's birth (in 1917). In one of the most brilliant short takes on *Walden* ever penned, she writes that the book imparts:

> a sense of beholding life through a very powerful magnifying glass. To walk, to eat, to cut up logs, to read a little, to watch the bird on the bough, to cook one's dinner—all these occupations when scraped clean and felt afresh prove wonderfully large and bright. The common things are so strange, the usual sensations so astonishing that to confuse or waste them . . . [seems] an act of sacrilege.[25]

A more quintessential expression from Thoreau's own era illustrating simplicity ethics of Wuthnow's first type, ascetic moralism, is a little book of 1829 which was about to go into its thirty-third printing just as Thoreau put the finishing touches on *Walden:* a tract called *The American Frugal Housewife.* Lydia Maria Child, its author, knew Thoreau slightly. She was a prolific New England–born woman of letters on the periphery of the Transcendentalist circle, best remembered today as one of the minority of Yankee antislavery activists not only to condone but also explicitly to defend what was then called "amalgamation," or marriage across the color line: first white and American Indian; then white and black. But for advocates of voluntary simplicity, *The American Frugal Housewife* has been Child's most inspirational text, albeit less famous than *Walden.*[26] The two books resemble each other in their vigilant and passionate attachment to small things and moments in daily life that most people overlook. But Child concentrates less on the higher forms of self-fulfillment that frugality confers than on frugal efficiency as a moral imperative in and of itself, as in this opening assertion: "Nothing should be thrown away so long as it is possible to make any use of it, however trifling that use may be; and whatever the size of a family, every member should be employed either in earning or saving money."[27] She also commends thrift, efficiency, and living within one's means as pragmatic survival strategies, but the overriding justification is moral: it is the right way to live.

That said, this brace of antebellum simplicity advocates also show that ascetic and expressive moralism are better conceived not as diametric opposites but as coordinate dispositions, often intertwined. Beyond a point, Wuthnow's distinction breaks down. Child takes an obvious relish in proliferating a cornucopia of ingenious tidbits like the virtues of old suds as fertilizer or New England rum as healthier for your hair than store-bought Macassar oil, while Thoreau for his part makes a point of advocating vegetarianism as morally superior to eating meat. When Thoreau recalls rejecting the innocent gift of an unwanted doormat because "it is best to avoid the beginnings of evil,"[28] he is being droll, but also serious.

The same mixed picture holds for much, if not all, modern versions of voluntary simplicity. They tend to rest on a fusion between expressive and ascetic values—between the carrot and the stick. Gregg's manifesto, for instance, argues for "singleness of purpose," but to the end of redirecting imagination toward "new" and worthier desires that would "secure greater abundance of life in other directions."[29] Bill McKibben's *Deep Economy* begins with a jeremiad addressed to the developed world's ecological conscience, warning that the modern fetishization of economic growth will destroy the planet, but opens up into what he calls "a new utilitarianism" that promises restoration of body and spirit as well as a brighter future for the earth itself if enough of us resolve to go local in our foodways and civic commitments.[30]

This hopeful prospect calls to mind a third strand of voluntary simplicity ethics today, exemplified by British philosopher Kate Soper's proposal for a counter-consumerist movement that she calls "alternative hedonism." Soper starts from the premise that the culture of consumption is so intractably ingrained that "the model of the 'good life' is unlikely to be checked in the absence of a seductive alternative"—meaning that an alternative ethos cannot succeed by appealing merely "to altruistic compassion and environmental concern," or to "a 'simpler' and more 'natural' system of need satisfaction."[31] It must not only address but also appeal head on to "the more self-regarding gratifications of consuming," to "the sensual pleasures of consuming differently," and to the "intrinsic pleasures these afford."[32] With that in mind, she takes heart in the perception that an affluent lifestyle is "generating its own specific forms of disaffection," through the congestion, noise, and stress of shopping malls, or the sensory deprivations of climate-engineered environments in which contemporary work, travel, and shopping increasingly take place, or the substitution of costly and unappetizing alternatives like treadmill gyms for the "straightforward and inexpensive pleasures" of walking or running in an increasingly urbanized outdoors.[33] An earlier, longer, coauthored version of this manifesto lays out a series of contrarian strategies for maximizing pleasure, health, and well-being for every dollar spent, beginning with "heritage cookery," as the authors call it.[34]

The alternative hedonism hypothesis makes a good vantage point for calling attention to certain question-begging aspects of voluntary simplicity ethics in whatever form. I want to take up three in particular, after which I shall return by way of conclusion to the question posed by my title: whether the example of Thoreau has any continuing use-value in the here and now.

The first question has to do with voluntary simplicity and *class*, or more precisely, level of affluence. Voluntary simplicity has generally been thought of as directed, like the Gospel of Luke in the Christian New Testament, toward "the privileged" rather than toward the poor.[35] Sociologist Amitai Etzioni puts the idea in starkest form: Voluntary simplicity is a choice faced by "a successful corporate lawyer," "not a homeless person" and by "Singapore, not Rwanda."[36] Though there is a ring of self-evident plausibility to this, it does not provide much help in drawing the line for those well off enough to warrant being made to feel guilty about not simplifying. The well-meaning exemption of the poor from the mandate is also somewhat dismissive. Lydia Child and Henry Thoreau lived in the days when moralistic distinctions were drawn between the deserving and the undeserving poor, before more compassionate and sociologically nuanced conceptions of charity displaced the earlier assumption that put the onus for rising from poverty on the individual. Indeed both of these writers sometimes voice this archaic ethos in quite repellent ways, as when Thoreau tries to impress upon a bewildered immigrant laboring family the contrast between his thrift and their improvidence in shelling out for such luxuries as tea and coffee, then throws up his hands, lamenting that "the culture of an Irishman is an enterprise to be undertaken with a sort of moral bog hoe."[37] Yet it is also refreshing and praiseworthy that both Thoreau and Child address their respective gospels of simplicity not only or even mainly to the wealthy but, even more importantly in their own eyes, to those hard pressed to make ends meet, who must figure out better ways (to recycle an old New England saying) of how to live off lack of expense.

The second question, related to the first, has to do with the *incrementalism* of voluntary simplicity ethics. Although voluntary simplicity defines itself as a countercultural movement, and in principle I agree that it is, the exemplary forms of behavior that it offers mostly fall into the category of such minor adaptations as composting garbage, recycling waste, opting for public transportation rather than personal auto, and the like. Prominent voluntary simplicity advocate Duane Elgin cautions that "it seems better to move slowly and maintain a depth of commitment that can be sustained over the long haul" than try to change your ways abruptly. Then, "gradually, a person or family may find they have made a number of small changes and acquired a number of slightly different patterns of perception and behavior, the sum total of which adds up to a significant departure from the industrial way of life."[38] Perhaps so. But one cannot help wondering how far the net result of acting upon such counsel is likely to get one beyond the kind of advice that Elaine St. James gives in *The Simplicity Reader* (1999), a trilogy of bestsellers each of which takes the form of one hundred tips on "ways to slow down and enjoy the things that really matter," such as "Go for patterned carpets," "Don't answer the phone just because it's ringing," "Break your routine once in awhile," and "Don't even think about saving that piece of aluminum foil."[39] In fairness, she also includes some much more high-stakes advice, like "Get rid of your car," and "Practice dying,"[40] but Albert Schweitzer or even Henry Thoreau this writer clearly is not. *The Simplicity Reader* mostly offers the practical face of Soper's alternative hedonism theory, divested of its postcapitalist theoretical framework: strategies for negotiating your way through consumer culture without forsaking it.

That may help explain why Duane Elgin both invokes Thoreau as a precedent for the modern movement and tries to keep him at arm's length: to dispel what Elgin grants is likely the novice's standard image of what voluntary simplicity must mean: namely, "a hardy person or couple who have turned away from material progress, moved to a rural setting, and chosen a life of isolated and austere simplicity."[41] Elgin does give some plausible reasons for this misperception, such as the stumbling block of imagining voluntary simplicity as possible only in an outback setting, or in isolation from other people. But the more fundamental challenge that *Walden*'s back-to-the-cabin experiment would seem to pose for small-step incrementalists is the degree to which it at least professes to break from mainstream habits. However much Thoreau's book is invoked honorifically as an ancestor, and despite Thoreau's explicit insistence that he does not expect or want others merely to copycat his example, inevitably he looks like an extremist to the folks who post such messages to today's sundry voluntary simplicity blog sites as "True frugality isn't just about spending less money," and "Eliminating everything except necessities . . . sounds pretty grim," to cite two recent postings on choosingvoluntarysimplicity.com.[42]

As seasoned Thoreauvians know, he himself was more of an incrementalist than a cursory reading of *Walden* suggests. Around the edges of the text, he admits to going back to town almost every day, to dining at friends' houses and hanging out in the village, and to completing the book after becoming "a sojourner in civilized life again."[43] One biographer sums up the Walden experiment in Eriksonian terms: it was intended not as a permanent withdrawal but as a "moratorium."[44] And as an earlier biography—by one of Thoreau's closest friends—states the case, his experiment in voluntary simplicity was designedly a near-home "bivouac" that allowed him to keep his social network intact, rather than a faraway removal.[45] Certainly Thoreau risked far less than, say, Oregonian ex-academic Charles Gray, who decided in 1977 at the age of fifty-two that "the mega-crisis of late Twentieth century humanity" demanded that he live within what he calls the "World Equity Budget"—his proportional one-person share of "the world's total [dollar] income," which he reckoned to be $142 per month at the time.[46] This decision immediately cost him his marriage and plunged him into severe loneliness and depression until he found ways to adjust.

Yet *Walden*'s staying power may have proven as great as it has precisely because Thoreau's experiment was both much more than a faint gesture *and* a controlled experiment kept within bounds. The scores of cases I have collected over the years of people who self-consciously tried to follow in his steps range from the tepid to the drastic, from suburban weekend warriors to bona fide hermit types much more extreme than he.[47] A recent write-up that caught my eye as an arresting metropolitan, media-age experiment but still of a legitimately neo-Thoreauvian stripe is Colin Beavan's *No Impact Man*, a book developed concurrently with a blog which the author still maintains. A forty-something couple with an infant daughter undertake a one-year experiment in eco-responsible scaling-down from their Manhattan apartment base that involves a series of increasingly drastic curtailments: first take the stairs, not the elevator, and walk, don't drive; then reduce trash to zero, including no more Pampers and styrofoam; then eat only food grown within 250 miles; then reduce electricity use to the minimum and depend to the max on a solar energy gadget on the roof, and so on. Despite a considerable degree of offputtingly narcissistic self-promotionalism (although that too is in a certain sense Thoreauvian), Beavan's story impressed me as an absorbing narrative of largely successful coping with an increasingly rigorous series of self-imposed constraints in an emphatically contemporary setting and as a compelling interweave of all three semirelated ethical strands I have identified: ascetic moralism, expressive moralism, and alternative hedonism. Roughly speaking, the book argues that if you accept the discipline of the first using the creativity of the second, the pleasurable compensations of the third will follow.[48]

My third question about voluntary simplicity is the potentially atomizing effect of its appeal first and foremost to the conscience of the individual. Although Elgin criticizes Thoreau for reclusiveness and emphasizes instead—as McKibben also does—the importance of voluntary simplicity ethics as a potential strengthener of the bonds of community, by his own admission, no organized collective voluntary simplicity movement as such exists. Voluntary simplicity is rather to be understood as a concatenation of voices and choices; the emphasis Elgin himself places on legitimate variation of personal choice seems systemically to prioritize individual action above united front.

On *this* issue, Thoreau provides no *ostensible* help at all. On the contrary, he seems, if anything, to offer an exaggerated case of voluntary simplicity's reform-oneself-first approach. *Walden* rails against philanthropy as officious do-gooderism. Thoreau's great political essay "Civil Disobedience" attaches preeminent importance to the heroic *individual* act of conscience in disentangling the self from the state.[49] It took Gandhi and King to convert Thoreau's vision into a collective movement with a wholeheartedly altruistic face. But at the same time, as Gandhi's conception of *satyagraha* stipulates, there is arguably something to the idea that "Union is only possible to those who are units," as Thoreau's fellow Transcendentalist Margaret Fuller once wrote.[50] The long-term influence of Thoreau's essay, as well as *Walden*, demonstrates the infectious potential of individual initiative expressing itself with conviction and eloquence. If the hyperindividualism that kept Thoreau from becoming a Gandhi is a trap, so too is waiting for a critical mass to form before you try to do anything. Colin Beavan, the would-be No Impact Man, makes a shrewd comment to this purpose as his parting shot:

> For most of my forty-five years . . . I [was] too paralyzed by this question of whether I was the type of person who could make a difference. Finally, during the year of the project, I realized that's the wrong question. The real question is whether I'm the type of person who wants to try.[51]

Still, the question remains as to whether voluntary simplicity at the individual (or in this case, family) level *will* generate any sort of broader impact. No Impact Man claims he has, that even after the year of the experiment ended: "I continue to change the people around me. We can all change the people around us by changing ourselves."[52] Maybe so. But there is no built-in guarantee that downshifters will deliver anything back to others. In addition to ascetic moralism, expressive moralism, and alternative hedonism, voluntary simplicity seems to stand in need of some other ingredient, some altruism clause, to maximize the likelihood that scaling-down would mean giving away as well as giving up. Charles Gray reports that whatever he earns above the $142 limit of the World Economic Budget he gives away. That is a small-scale version, but to my mind equally admirable, of the Australian millionaire Karl Rabeder, who recently announced that he planned to donate his entire fortune to a microcredit nonprofit he had set up for small entrepreneurs in Latin America, and "move into a small wooden hut in the mountains or a studio in Innsbruck."[53] The Buell household's far more modest redistribution scheme is to give to charity each year an amount at least equal to what we spend on entertainment or recreation of any kind.

But the last word on this subject should rightfully go to a much more heroic exemplar of voluntary simplicity, Albert Schweitzer. Whatever might be said against Schweitzer—and much can—for Eurocentric paternalism toward Africans as virtual children, his 1931 autobiography, translated as *Out of My Life and Thought*, remains a classic testimonial of one who opted for a life of relative personal privation in the service of others. The chapter in which he recounts his decision to become a medical missionary is remarkable both for self-exacting admonitions to the naively impressionable reader that very few people should undertake such a drastic life change as he did, and for the importance attached to "the hidden forces of goodness . . . embodied in those persons who carry on as a secondary pursuit the . . . service which they cannot make their lifework" for whatever reason. On the practice of personal altruism in such comparatively quotidian ways "depends," Schweitzer insists, "the future of mankind," convinced as he is that "our humanity is by no means so materialistic as foolish talk is continually asserting it to be."[54] In Schweitzer, then, we find a figure whose own career would seem to embody the kind of heroic renunciation of conventional success paths that, on the face of it, makes the ethics of voluntary simplicity look timid and puny, affirming the significance of modest and unobtrusive acts of generosity on the part of those who stop short of committing to drastic life changes.

To be sure, Schweitzer wrote this more than seventy-five years ago, before World War II, before the Holocaust, before the nuclear age, before Vietnam, before Rachel Carson, before 9/11, before the discovery of anthropogenic climate change. Does the liberal individual-centric model on which contemporary voluntary simplicity ethics is based, and which in a sense Schweitzer presupposes, still have traction today? Were we to call Schweitzer back from the grave, and Thoreau too for that matter, I should not be surprised if both would continue to affirm it as necessary, if not sufficient—necessary as a precondition for human integrity and flourishing, even if not sufficient to preserve the world from ruin in the long run. If that is right, then humankind would be foolish to trust to voluntary simplicity alone but equally foolish to disown it.

Notes

1. I am most grateful for many constructive suggestions made during and after the lecture presentation of this talk, especially by Diana Eck's wise and searching commentary on the lecture version. A few of these I have incorporated here, although because Professor Eck's commentary also appears in this volume I refrain from revising as much as I am tempted, in order to remain faithful to the main contours of the original text. I would also like to express special gratitude to Donald Swearer—close colleague and friend for more than forty years, to whom I dedicate this essay. Emerson's resonant aphorism, "Nor knowest thou what argument thy life to thy neighbor's creed has lent," was never better embodied than in the example Don has set for all those whom he has influenced.
2. Samuel Alexander, ed., *Voluntary Simplicity: The Poetic Alternative to Consumer Culture* (Whangangi, NZ: Stead & Daughters, 2009).
3. Ralph Waldo Emerson, *Essays, First Series* (1841), ed. Joseph Slater et al. (Cambridge: Harvard University Press, 1979), 131.
4. Alan Thein During, *How Much Is Enough? The Consumer Society and the Future of the Earth* (London: Earthscan, 1992), 28.
5. Wendell Berry, *A Continuous Harmony: Essays Cultural and Agricultural* (San Diego: Harcourt, 1972), 74.
6. The survey information is taken from United Nations Environment Programme (UNEP), *Global Environment Outlook (GEO)-1, Global State of the Environment Report* (Nairobi, Kenya: UNEP, 1997), available in book form and on the web at http://www.grida.no/publications/other/geo1/. The material referenced here is from the online version, from "Underlying Causes," in the North America section of chapter 2: Regional Perspectives. The Merck Family Fund commissioned the original survey and report; see Merck Family Fund, "Yearning for Balance," prepared by the Harwood Group (Takoma Park, MD: Merck, July 1995), available online at http://www.iisd.ca/consume/harwood.html.
7. David E. Shi, *The Simple Life: Plain Living and High Thinking in American Culture* (New York: Oxford University Press, 1985), 277–279.
8. While I was composing this essay, however, after what seems to have been a highly successful tenure, President Shi announced his forthcoming resignation and his intent to return to his first love, history.
9. Lizbeth Cohen, *A Consumer's Republic: The Politics of Mass Consumption in Postwar America* (New York: Knopf, 2003), 118, 127.
10. Kate Soper, "Alternative Hedonism, Cultural Theory and the Role of Aesthetic Revisioning," *Cultural Studies* 22 (September 2008): 568, 401.
11. Richard Layard, *Happiness: Lessons from a New Science* (New York: Penguin, 2005), 230. See also Daniel Akst, "A Talk with Betsey Stevenson and Justin Wolfers: It turns out money really does buy happiness. Uh-oh," *Boston Globe*, November 23, 2008, and Betsey Stevenson and Justin Wolfers, "Economic Growth and Subjective Well-Being: Reassessing the Easterlin Paradox," *Brookings Papers on Economic Activity* (Spring 2008): 1–87, doi: 10.1353/eca.0.0001.

12. Layard, *Happiness*, 41.
13. Robert M. Biswas-Diener, "Material Wealth and Subjective Well-Being," in *The Science of Subjective Well-Being*, ed. Michael Eid and Randy J. Larsen (New York: Guilford, 2008), 314.
14. Stevenson and Wolfers, "Economic Growth and Subjective Well-Being," 58–87.
15. Richard B. Gregg, *The Value of Voluntary Simplicity* (Wallingford, PA: Pendle Hill, 1936), 4–5, 25.
16. Ibid., 3, 23.
17. Ralph Waldo Emerson, "Thoreau" (1883), in *Complete Works of Ralph Waldo Emerson: Lectures and Biographical Sketches*, ed. Edward Waldo Emerson (Boston: Houghton, 1903–1904), 479.
18. Robert Wuthnow, *Poor Richard's Principle: Recovering the American Dream Through the Moral Dimension of Work, Business, and Money* (Princeton: Princeton University Press, 1996), 340.
19. Ibid., 72.
20. Emerson, "Thoreau," 52.
21. Henry David Thoreau, *Walden; or, Life in the Woods* (1854), ed. J. Lyndon Shanley (Princeton: Princeton University Press, 1971), 70.
22. Ibid., 15.
23. Ibid., 89–90.
24. Shirley, "Downshifting to a Simpler Life," Choosing Voluntary Simplicity (blog), http://www.choosingvoluntarysimplicity.com.
25. Virginia Woolf, "Thoreau," in *Times Literary Supplement*, July 12, 1917. Reprinted in *Books and Portraits*, ed. Mary Lyon (London: Hogarth Press, 1977), 75.
26. Lauren Weber, *In Cheap We Trust: The Story of a Misunderstood American Virtue* (New York: Little, 2009), 86–87.
27. Lydia Maria Child, *The American Frugal Housewife* (1832; Project Gutenberg, 2004, eBook no. 13493), 5.
28. Thoreau, *Walden*, 67.
29. Gregg, *Voluntary Simplicity*, 4, 28.
30. Bill McKibben, *Deep Economy: The Wealth of Communities and the Durable Future* (New York: Times Books/Holt, 2007), 45.
31. Soper, "Alternative Hedonism," quotes in order: 571, 574.
32. Ibid., 572.
33. Ibid., 577.
34. Kate Soper and Lyn Thomas, "'Alternative Hedonism' and the Critique of 'Consumerism,'" *Cultures of Consumption* Working Paper 31 (London: Birkbeck College, December 2006), 12–44. Available online at http://www.consume.bbk.ac.uk/publications.html.
35. Gregg, *Voluntary Simplicity*, 30.
36. Amitai Etzioni, "Voluntary Simplicity: Characterization, Select Psychological Implications, and Societal Consequences," *Journal of Economic Psychology* 19 (1998): 632. See also Duane Elgin, *Voluntary Simplicity: Toward a Way of Life That Is Outwardly Simple, Inwardly Rich* (New York: Morrow, 1981), 47.

37. Thoreau, *Walden*, 205–206.
38. Elgin, *Voluntary Simplicity,* 57.
39. Elaine St. James, *The Simplicity Reader: Simplify Your Life* (vol. 1), *Inner Simplicity* (vol. 2), and *Living the Simple Life* (vol. 3) (New York: Smithmark Press, 1999), 1:31, 76; 2:346; 3:667.
40. Ibid., 1:230, 2:371.
41. Elgin, *Voluntary Simplicity,* 20, 37.
42. Shirley, "What True Frugality Is . . ." and "What Voluntary Simplicity Is NOT," *Choosing Voluntary Simplicity* (blog).
43. Thoreau, *Walden*, 3.
44. Richard Lebeaux, *Young Man Thoreau* (Amherst: University of Massachusetts Press, 1977), 216–217.
45. William Ellery Channing, *Thoreau the Poet-Naturalist,* ed. Franklin B. Sanborn (Boston: Goodspeed, 1902), 24. *Walden* itself actually acknowledges this more openly than casual readers tend to notice, beginning with Thoreau's introduction to the detailed description of his experiment: "It would be of some advantage to live a primitive and frontier life, though in the midst of an outward civilization, if only to learn what are the gross necessaries of life and what methods have been taken to obtain them." Thoreau, *Walden,* 11.
46. Charles Gray, "The World Equity Budget or Living on about $142 per Month," in *Downwardly Mobile for Conscience's Sake: Ten Autobiographical Sketches,* ed. Dorothy Anderson (Eugene, OR: Vesta, 1995), 98–100.
47. Lawrence Buell, *The Environmental Imagination: Thoreau, Nature Writing, and the Formation of American Culture* (Cambridge: Harvard University Press, 1995), 325–327.
48. Colin Beavan, *No Impact Man: The Adventures of a Guilty Liberal Who Attempts to Save the Planet and the Discoveries He Makes about Himself and Our Way of Life in the Process* (New York: Farrar, Straus, 2009).
49. Henry David Thoreau, "Resistance to Civil Government" [Thoreau's original title], *Reform Papers,* ed. Wendell Glick (Princeton: Princeton University Press, 1973), 63-90.
50. Margaret Fuller, *Woman in the Nineteenth Century*, ed. Larry J. Reynolds (New York: Norton, 1998), 71.
51. Beavan, *No Impact Man*, 224.
52. Ibid.
53. Terence Neilan, "Millionaire Is Giving Away His Entire Fortune," 14 February 2010, http://www.aolnews.com.
54. Albert Schweitzer, *Out of My Life and Thought* (1931), trans. C. T. Campion (New York: Mentor, 1949), 76, 77.

Voluntary Simplicity: Thoreau and Gandhi
Response to Buell's *Does Thoreau Have a Future?*

Diana Eck

Walden is one of the few works of American literature that comes close, I think, to having the status of scripture. When Wilfred Cantwell Smith, the eminent historian of religion, asked "What is scripture?" he concluded after comparative study that "scripture" is not the text itself, but the particular, special relationship that we have with the text as individuals and as a community.[1] Like many others, I have that special relationship with *Walden*. It is one of the few books with which I have that kind of relationship, although I am a Bible-reading, Bible-studying Christian and I also read, study, and love the *Bhagavad Gita* and *Upanishads*. But *Walden* is the book that stays by my bedside. I can pick it up randomly and read for five minutes, morning or evening, and find something quite wonderful: "Let us spend one day as deliberately as Nature"; let us "not be thrown off the track by every nutshell and mosquito's wing that falls on the rails"; or "Let us rise early and fast, or break fast gently and without perturbation."[2] Written carefully, *Walden* asks us to read it as carefully as it was written. Created out of an experience of voluntary simplicity, a short-lived experiment, it became a model for other experiments, including that of Colin Beavan's *No Impact Man*, which Professor Buell describes in the preceding essay.

For me, this "scripture" of Thoreau provides not so much a model as an inspiration, a provocation, a call to awaken. Who can forget, for example, those three pieces of limestone that he had placed on his desk for their simple beauty, only to realize that they needed dusting? Thoreau felt he did not have time to dust the furniture of his own mind, let alone those three pieces of stone on his desk—so out they went.[3] Elsewhere, he comments on the fascination with houses we human beings seem to have: "At a certain season of our life we are accustomed to consider every spot as the possible site of a house."[4] And then there is the matter of furnishing the house: "I had three chairs in my house; one for solitude, two for friendship, three for society."[5] No matter where I open *Walden* and start reading, I find nourishment and something to contemplate. As long as we are attuned to and willing to be provoked by that ornery ancestor at Walden, his inspiration will live on. *Walden* asks questions that we need to ask on a regular basis, whether or not we are bold enough to change our lives in response.

Why was voluntary simplicity important for Thoreau? Perhaps more significantly, why is it important for us as we reimagine voluntary simplicity in the twenty-first century? Thoreau stands in the American tradition of deliberately setting out into nature, to carve something out with the rudiments of bare hands and hard work. My own great-grandparents homesteaded on the frontier as subsistence farmers in the Pacific Northwest, in the hills of the Olympic Peninsula where farming was hard and the season short. My great-grandfather did not take to the farming life, so he would go into the woods to shoot a bear or a couple of elk or deer to feed his family. My great-grandmother did a little garden-patch farming. For a living, she picked blackberries in July and August, with her holster and pistol on one hip, a child on the other, and a blackberry bucket over her arm. She would exchange blackberries for what she needed down at the store in Sequim. They chose this life. Indeed, they moved out of a fine little town, Port Townsend, to go up there and live deliberately. In what ways were their lives different from the ethics described in the voluntary simplicity literature that we have today? Larry Buell offers us Richard B. Gregg's definition: "Voluntary simplicity means an ordering and guiding of our energy and our desires. A partial restraint in some directions in order to secure greater abundance of life in other directions."[6] My great-grandparents were restraining themselves in one direction. Without idealizing what was surely a difficult life, there is something beautifully abundant in my great-grandmother's account of that choice: "The cabin was 12 by 14 feet. The branches from the fir trees, chopped fine, made a good covering for the floor and the smell of balsam was soothing." And she left us a prayer from those years: "Grant us the peace of solitude. Give us the strength of the forest. For you know, O Lord, that we love not our fellow beings less, but nature more."[7]

"To secure abundance in other directions" brings to mind two of my younger colleagues. The first, Rebecca Gould, teaches now at Middlebury College; among her course offerings is "The Simple Life in America." In her book, *At Home in Nature,* she looks at some of those who decided deliberately in the late-twentieth century to homestead out in the country. She researched Helen and Scott Nearing, Wendell Berry and his writings, and other contemporary votaries of the simple life. She interviewed many of these modern pioneers who see homesteading as a form of dissent from contemporary consumer culture, a departure from a traditional religious life, and also a practice of environmental ethics resembling a religion in itself. In the book, Gould examines the parallels in the lives of these modern homesteaders and thinks about them in relation to Thoreau as well as suggesting what their choices offer us as an ethos of life for today.[8]

The other colleague I want to cite is Kimerer LaMothe, former Director of Undergraduate Studies at the Committee on the Study of Religion at Harvard. Kimerer is a theologian and a dancer. She studies the body and dance as a way of thinking about religion. Eventually, Kimerer and her musician husband left the urban environment of Cambridge with their three—now five—children to resuscitate a rundown farm in rural New York. Farming is a hard life and not necessarily a simple one, but it is connected to the rhythms of the natural world. This sense of connection is what LaMothe and her husband wanted to provide their children. As she writes in the manuscript of her new book, *Family Planting,* she has found Thoreau an inspiration in her quest:

> He was dogged by a nagging suspicion that the urban ladder fostered ways of thinking, and feeling, and acting that were not healthy. We were, too. Like Thoreau, I wanted to align my art into a sense of dance more deliberately with the forces that sustained the best in human being. And for a family, the lesson of living, and loving, and working with nature in a mutually enabling life, is an incredible experiment, and a difficult one. Thoreau was seeking a closer relationship to the natural world. Not only as a refuge from

> the noise of social life, but also as a fund of patterns and symbols that would help him understand and live his life fully awake and with love.[9]

LaMothe's remarkable book explores the relationships of love in the midst of these patterns and symbols. Here, she has to depart somewhat from Thoreau's voluntary simplicity, for these relations of family love and the bonds of connection that are stretched and strengthened through the years are never simple.

In surveying the landscape of voluntary simplicity in the essay preceding this one, Buell raises three significant questions to which I would like to return. First, is voluntary simplicity only a class issue, voluntary because only the rich can afford it? If you have too much, it is attractive to think about giving up some of the clutter of this life. Who of us would not want to throw out, not only the three limestone pieces, but half of the items in our homes to gain more simplicity? Is this the corporate lawyer rather than a homeless person? Yes. Voluntary simplicity may be oriented to the class of those who have, rather than those who do not, but that gap is at the heart of the issues of economic justice in our society. The advocates of voluntary simplicity that I know best, Gandhi and his disciple Vinoba Bhave, walked hundreds and hundreds of miles through the villages of India talking about "trusteeship," encouraging a form of voluntary simplicity. They carried the message that the land and property we have is ultimately not ours, but only ours in trust for everyone else in society. Through the Bhoodan (land-gift) movement, Vinoba persuaded people with land to give some of it voluntarily to those without land. Eventually some five million acres were gifted in this way. Yes, that is a class issue and it is also a movement of voluntary simplicity at its best.[10]

Second, is voluntary simplicity only about making small adjustments? A type of incrementalism? Today there are many small adjustments being made all around us, even here at Harvard. The students recently started a small organic gardening project in a lot right outside the main entrance to Lowell House, the student residence of which I am House Master. The raised garden beds have been flourishing, although the skeptic could complain that most of what they produce in a season would be consumed in Lowell House dining hall in a single day. The dining halls have instituted voluntary meatless Mondays. Recycling and taking account of what we waste is another incremental adjustment that has become part of the consciousness of students and many others in our society. So do small adjustments make any difference at all in the midst of the maelstrom of commercial life and mass production? As an educational matter, I think they do. I think Gandhi, along with Thoreau, would also look at these experiments in gardening, recycling, and all the other small incremental actions, with a nod of satisfaction. Small as they may seem, this is where revolutions begin. If they do not begin here, they begin nowhere.

We know from the example of the pioneers of voluntary simplicity that small steps help us to rehabituate ourselves. Why would Gandhi use a spinning wheel? What would be the point of hand-spinning cotton in this day and age? Isn't that simply a bit of symbolic incrementalism? But spinning became the act that even a villager could undertake in order to participate in the movement for self-rule, *swaraj*. It became the emblem of empowerment, when individuals felt powerless to make a difference.

Gandhi's first *satyagraha* campaigns in India responded to the economic issues of the day in ways that must have seemed like symbolic gestures, at least to his critics. *Satyagraha*, literally "holding fast to truth," was Gandhi's term for nonviolent action. Who would have imagined that his first campaign on behalf of suffering indigo farmers in a remote part of India would bring him national attention? And what about his action protesting the government monopoly on salt? This too was a seemingly small symbolic issue that quickly focused a nationwide campaign. Gandhi's Salt March, his two-hundred-mile pilgrimage by foot to the sea in defiance of the government monopoly on the production of salt, was a brilliant move. Harvesting salt from the sea created a powerful movement, defying the laws of colonial India on an issue that touched the lives of everyone. These symbolic economic issues were an intrinsic part of the Gandhian strategy and ethic.[11]

The third question: is voluntary simplicity too individualistic? Will it ever make a difference more broadly? I loved the quote from *No Impact Man* that Buell cites in which Beavan says the question is not whether he will ever make a difference in the world, but will he try.[12] Individual action is the first and foremost difference that each of us can make. Gandhi, Thoreau, and Tolstoy all tried to make a difference. They each started individually. Gandhi took his individual ethic and brought it to community, first by creating communities in South Africa and then in India, and second by being an individual example to awaken and inspire others. We all know the Gandhian dictum, "Be the change you want to see in the world."

Voluntary simplicity addresses the problem of habitual living. It helps to bring into our consciousness, our ethics, and our religious life an ability to reflect more deeply on how we live, reminding us to choose what we consume, instead of being swept along by the tremendous force of our habits and of the capitalist consumerist society in which we live.

As a way of being in the world, consumerism affects the poor as well as the wealthy. It is not just an issue for the rich who have fat wallets. The cycle of desire and acquisition feeds on itself, creating and fulfilling demand. It is a cycle in which desire and demand are never satisfied. This relentless pattern of consumerism is a habitual way of thinking that is not simply an economic system, but a deeply ingrained way of living.

At the heart of consumerism is the most intimate consumer issue of all—what we eat. What items do we choose to put in our mouths every single day? Where do they come from, who makes them, how many of them do we eat? Thoreau has some wonderful meditations on food, but I want to turn again to Gandhi. He also paid scrupulous attention to food. People who knew him talked about his seemingly eccentric food habits, for he believed that what we eat and how much we eat is a deeply ethical issue. He was always experimenting with diet in relation to health as well, giving up cow's milk for goat's milk, tea for cocoa. Gandhi ultimately decided that he would have no more than five food items a day—not because of any intrinsic virtue in that number, but to set a limit to this first and most important of consumer issues.[13] This was voluntary simplicity in the realm of one's diet, and in a culture of food and flavor, this is a radical act of voluntary simplicity.

Consumerism is linked to exploitative economic injustice and unfair labor practices. In Gandhi's little book *Trusteeship* he suggests that we are all thieves insofar as we aggrandize our desires and accumulate far more than we need. He writes, "If I take anything I do not need for my own immediate use, and keep it, I thieve it from someone else. I venture to suggest that it is the fundamental law of Nature, without exception, that Nature produces enough for our wants from day-to-day, and if only everybody took enough for himself and nothing more, there would be no pauperism in this world, there would be no man dying of starvation in this world."[14]

In *The Kingdom of God Is Within You,* Tolstoy makes a similar critique, reasoning from Christian principles:

> We are all brothers, yet every morning my brother or my sister carries out my chamber pot. We are all brothers, and I need every morning my cigar, sugar, a mirror, so forth, objects in the manufacture of which my brothers and sisters, who are my equals, have been losing their health, and I employ these articles, and even demand them. . . . We are all brothers, and yet I live by receiving a salary for arraigning, judging, and punishing a thief or a prostitute, whose existence is conditioned by the whole composition of my life.[15]

Like Tolstoy, Gandhi saw that the economic issues involved in habitual consumerism profoundly undergird our daily lives, especially when we depend on food and goods from far away, made often at the expense of exploiting our brothers and sisters. Both Thoreau and Gandhi were concerned about the local. In economic terms, they were concerned about *swadeshi*, the practice of using those things that are made close to one's own *desh* (place of residence). They promoted the spirit that gives priority to one's immediate neighbors and uses local products, rather than imports. In Gandhi's manifesto *Hind Swaraj* (*Indian Home Rule*) he writes that political "home rule" must begin literally at home, in one's immediate domestic environment and in the willingness to do for oneself what has too often been provided through conditions of exploitation.[16] Gandhi would have understood Thoreau's small-scale domesticity, his cultivation of beans, and his growing self-reliance. He would also have understood what Thoreau meant about those three pieces of limestone. He himself was famous for how few possessions he owned at the end of his life.

Larry Buell has explored the idea that voluntary simplicity might need an altruism clause to keep it from being simply narcissistic and self-centered. In talking about these exemplars of voluntary simplicity, I am suggesting that it already contains that altruism clause. Voluntary simplicity asks us to consider our ethics and the way we live, with each other and in relation to the environment. In Thoreau's words, it asks us to "live deliberately" and to consider others. It asks us to wake up.

Notes

1. Wilfred Cantwell Smith, *What Is Scripture? A Comparative Approach* (1993, repr. Minneapolis: Augsburg Fortress Press, 2005).
2. Henry David Thoreau, *Walden; or, Life in the Woods* (1854), ed. J. Lyndon Shanley (Princeton: Princeton University Press, 1971), 97, quotes in order.
3. Ibid., 36.
4. Ibid., 81.
5. Ibid., 140.
6. Richard B. Gregg, *The Value of Voluntary Simplicity* (Wallingford: Pendle Hill, 1936), 4–5. as quoted in Lawrence Buell's essay in this book; see Buell's note 15.
7. Hilda Fritz diary, papers in the author's possession.
8. Rebecca Kneale Gould, *At Home in Nature: Modern Homesteading and Spiritual Practice in America* (Berkeley: University of California Press, 2005).
9. Kimerer LaMothe, early manuscript of *Family Planting: A Farm-Fed Philosophy of Human Relations* (Ropley, United Kingdom: O-Books, forthcoming), http://www.o-books.com/book/detail/1143/Family-Planting.
10. For the Bhoodan movement, see Mark Shepard, *Gandhi Today: A Report on Mahatma Gandhi's Successors* (Arcata, CA: Simple Productions, 1987). Chapter 1, "The Saint and the Socialist," covers Vinoba Bhave and Jayprakash Narayan. See also Subhash Mehta, *The Bhoodan-Gramdam Movement–50 Years: A Review,* online at the website of the Gandhi Museum in New Delhi, http://www.gandhimuseum.org/sarvodaya/vinoba/bhoodan.htm.
11. See M. K. Gandhi, *An Autobiography, or, The Story of My Experiments with Truth,* trans. Mahedev Desai (Ahmedabad: Navajivan Pub. House, 1996), for his perspective on *satyagraha* (throughout, but particularly 266), homespun clothing (particularly 407–414), and indigo (337). See Yogesh Chadha, *Gandhi: A Life* (New York: Wiley, 1998), 291–298, for information on the Salt March; the book also provides perspective on *satyagraha.* For a fairly extensive overview of Gandhi's life and writings, see Homer A. Jack, *The Gandhi Reader: A Sourcebook of His Life and Writings* (1956, rev. ed., New York: Grove Press, 1994).
12. Colin Beavan, *No Impact Man: The Adventures of a Guilty Liberal Who Attempts to Save the Planet and the Discoveries He Makes about Himself and Our Way of Life in the Process* (New York: Farrar, Straus, 2009), 224, as cited by Buell in note 51 of the previous essay.
13. For some of Gandhi's thoughts on food, see Gandhi, *An Autobiography,* 47–50 and 267–278. For more on Gandhi's five food items a day, see Judith Margaret Brown, *Gandhi: Prisoner of Hope* (New Haven: Yale University Press paperback, 1991), 98.
14. M. K. Gandhi, *Trusteeship,* compiled by Ravindra Kelekar (Ahmedabad: Navajivan Publishing House, 2004), 1.
15. Leo Tolstoy, *The Kingdom of God Is Within You,* trans. Leo Weiner (repr. New York: Cosimo, 2007), V, 122–123. Available online at The Picket Line: http://sniggle.net/Experiment/index.php?entry=tkogiwy#C01.
16. M. K. Gandhi, *Hind Swaraj and Other Writings,* ed. Anthony J. Parel (Cambridge: Cambridge University Press, 1997).

Ecologies of Human Flourishing: A Case from Precolonial South India

Anne E. Monius

Not surprisingly, the essays in this book focus on the contemporary.[1] Phrases such as "world in crisis," "global climate change," "global health," "life on a tough new planet," and "the twenty-first century" are threaded throughout the book. The brochure for the lecture series from which this book sprang uses the language of contemporary "global crisis" demanding a radical transformation of our "'more is better' consumerist lifestyle."[2]

What possible contribution might a historian of religions working entirely on literary texts from precolonial South Asia make to this discussion? The potential relevance to contemporary and urgent global concerns is perhaps less than immediate. I would like to begin, in fact, by posing a prior question: What does it even mean to examine ecologies of human flourishing at times, places, and cultural locations profoundly distant from our own? Can it be done? Can the pieces of any particular "ecology" even be reconstructed? What, if anything, might be learned from such an exploration, as a historian of religion or as a citizen of the contemporary world? This is not the sort of question that South Asianists specializing in precolonial topics usually raise in public, nor does it allow me to stay entirely within the comfort zone of textual materials. With all the caveats that go along with wading into relatively unknown waters of art history, material culture, epigraphy, and agricultural technique, what follows is a historical case study of sorts, a brief look at one complex vision of human flourishing from a fascinating corner of precolonial South India: the Pallava dynasty that ruled from Kāñcīpuram (in the contemporary Indian state of Tamil Nadu) from roughly the seventh through ninth centuries CE.

Why the Pallavas?

I choose the Pallavas largely for two reasons, both because theirs is a courtly literary culture I have studied for some years and because of the evidence that their three (or more) centuries of rule yield for the topic at hand. In a region where the historical/textual record extends back at least to the second century BCE,[3] the Pallava polity is the first to emerge with anything resembling historical coherence in South India, in the northern part of the Tamiḻ-speaking region whose contemporary center lies in the bustling metropolitan center of Ceṉṉai. Pallava monarchs sponsored, oversaw, and even actively contributed to a complex literary culture spanning a wide variety of religious communities, languages, and forms. Indeed, Pallava rulers established a pattern in South India that would endure for a millennium: kings—whether Śaiva (devoted to the Hindu god Śiva) or Vaiṣṇava (devoted to the Hindu god Viṣṇu), Buddhist or Jain, Kannada-, Telugu-, or Tamiḻ-speaking—continually supported diverse religious institutions, communities, and literatures. Within the Pallava polity, local-language devotional literature—the poetry of the great *bhaktas* or saints—emerges for the first time in India among Śaivas and Vaiṣṇavas. Indeed, the primary religious institutions emblematic of the South even today—the massive stone Hindu temples dedicated to Śiva, Viṣṇu, or the goddess—are a Pallava innovation. The Pallava courts themselves, no less their cultural products, bespeak a continued effort at what one might call a mediation of sorts: between local and translocal, Sanskrit and Tamiḻ, northern and southern, South and Southeast Asia, Śiva and Viṣṇu, Jina and the Buddha.

As with all topics precolonial in South Asia, historical evidence of the Pallava rulers and their activities is admittedly fragmentary. As the late historian of South India, Burton Stein, once wrote, "extant historical evidence would probably not support the formulation of an ecotypology for any time earlier than the eighteenth century [in South India], and then only fragmentarily."[4] Yet in addition to their literary texts and institutions, the Pallavas are the first southern dynasty to leave behind a substantial body of inscriptions, offering a valuable glimpse at royal proclamation and the details of governance: some thirty grants written on copper plates and approximately two hundred inscriptions on stone by or about Pallava kings have been recovered from across South India.[5] Taking inspiration from Stein—who, despite the caveat above, goes on to outline an ecotypology of ritual polity, segmentary state, and peasant-Brahmin agrarian alliance for the later Cōḻa period (ninth through twelfth centuries) in South India—the following discussion attempts to sketch out a tentative ecotypology or ecology of human flourishing as envisioned and enacted by three centuries of Pallava rulers, with a particular emphasis on the projects of one especially powerful and productive king. In addition to the work of Stein, his students, and his detractors, this study is also inspired in part by the work on complex systems in Bali done by Stephen Lansing, where a much more fulsomely continuous historical record allows Lansing to postulate strong systemic connections among water management techniques, agricultural practices, social forms, and religious life.[6] While Pallava-era South India cannot yield the historical detail that both Lansing's and Stein's arguments require, simply raising the question of ecotypology, system, or ecology of human flourishing allows the historian to look at what evidence we do possess in new and potentially productive ways. While speculative and building on the work of many others, this foray into the visions of human flourishing among the Pallava rulers will perhaps modestly point the way to some new questions and strategies for reading the extant evidence.

The Pallavas of Early Medieval India

First, who exactly were the Pallavas of Kāñcīpuram? Theirs is the first dynastic lineage to emerge from multiple textual and material sources in southernmost India. Their origins—even after a century or more of scholarly speculation[7]—remain mysterious. They seemingly hail originally from somewhere other than the Tamiḻ-speaking region; the Pallavas are not included among the three great dynasties of the classical Caṅkam literary corpus (the Cōḻas, the Cēras, and the Pāṇṭiyaṉs) dating from the early centuries of the Common Era, nor does the long classical hymn of praise to the city (Kāñcīpuram) that would become their capital mention the name even once.[8] Even the proper succession of rulers' names has been a matter of some debate.[9] By the time the Pallavas emerge in a coherent-if-incomplete historical picture, the great religious transformations wrought by *bhakti* (devotion to a personal and loving god) are in full swing; the dynastic project of mediating land and sea trade, northern and southern cultural forms, and Sanskritic Vedic religious and cultural heritage with more local Tamil̲ idioms is already well under way.[10] The Pallavas are also great builders—of temples, palaces, monasteries, tanks, canals, and irrigation works—which themselves embody something of a mediation project. Earlier Tamil̲ poetic forms, for example—the above-mentioned classical or Caṅkam anthologies on the themes of love and war—focus on an idealized aesthetic of *tiṇai* or landscape; each poetic scene, whether of seaside, desert, mountain, forest, or agricultural tract, evokes for the educated reader a specific time, season, collection of flora and fauna, and, most importantly, a moment in the complex relationship between human lovers.[11] This pre-Pallava aesthetic is one of a natural world highly stylized, where the primary edifices are the huts of hunters and the modest homes of farmers.[12] Yet with the rise of a new Tamil̲ poetry of devotion to a personal and loving god during the reign of the Pallavas, the literary focus shifts quite dramatically to the importance of the *built* environment, to the palaces of kings and temples as the palatial homes of the gods on earth. "The town without a temple," sings the seventh-century devotee of Śiva, Appar, "is not a town. . . . It is a jungle!"[13] Unlike the wholly Tamil̲ aesthetic of the classical Caṅkam corpus, the built environment envisioned by this new Pallava-era literary outpouring blends together the literary, emotional, and ritual registers of both Tamil̲ and pan-Indic Sanskrit. "Behold the one who is northern Sanskrit, southern Tamil̲, and the four Vedas," Appar sings in another hymn, "behold Śiva, our treasure [enshrined] in Civapuram!"[14]

Pallava Water Imagery and Management

One fascinating and potentially relevant aspect of Pallava culture in all its forms—in literature, monument, and pragmatic construction—that few have taken note of is a consistent thematic interest in the management of water, inducing the monsoon to come in a timely and regular fashion, thus keeping the ever-looming specter of drought and famine at bay. In literary terms, this marks a substantial departure from the earlier managed landscapes of the classical poetic corpus, where each idealized region presents itself with climactic and emotional regularity: in the *mullai* (forest) landscape, for example, it is the rainy season of lovers in domestic happiness, while in the *pālai* (wasteland) lovers continually suffer the hardships of withering summer heat, deprivation, and separation.[15] While Pallava-era poets across religious communities build on this earlier landscape aesthetic in a variety of ways, unmistakable is a new—and very human—note of anxiety about the landscape run amok, about the unpredictability of the natural world and the profound human suffering that its vicissitudes can cause. Perhaps the earliest Pallava text to note what will become a consistent literary theme of the fickleness of the annual monsoon rains is the sixth- or seventh-century Buddhist *Maṇimēkalai*, the story of a young courtesan who foregoes her hereditary occupation to become a Buddhist nun. Maṇimēkalai enters the great city of Kāñcīpuram, but finds it devastated by drought. The king wonders aloud in her presence, before she begins to feed the starving multitudes from a miraculous begging bowl:

> I do not understand why this great land suffers poverty.
> Was it the straying of the royal scepter?
> Was it some error in performing austerities?
> Was there some imperfection in the virtue of the women
> whose hair is full of honey-laden blossoms?[16]

We will return to the king's musings in a moment. Such anxiety over the rains and the ever-present possibility of widespread hunger can be found throughout the literatures of Pallava-era South India, from Buddhist texts such as the *Maṇimēkalai* to Jain narratives and the poetry of the great *bhakta* saints. For the followers of Śiva, it is the lord Śiva himself who both wreaks havoc in the natural world and the human body and provides its cure: "We are subject to no one," sings Appar, "we do not fear Yama [the god of death]. . . . We do not know disease."[17] Cuntarar in the eighth century notes that at a temple of Śiva in Kuṇṭaiyūr, "I received some rice to stave off starvation,"[18] and praises his lord Śiva at Pukalūr as the father "who gives us boiled rice and clothing in this life."[19] Appar's younger contemporary, the child-saint Campantar, implores Śiva to grant him gold coins at Tiruvīḻimiḻalai that he might relieve the hunger of his followers during a time of great famine: "Bestow [on us] gold coins, oh lord of Mi<u>l</u>alai [a place-name]!"[20] Moving from Tami<u>l</u> to Sanskrit, the great court poet and literary theorist of the seventh century, Daṇḍin, narrates something of his own life trajectory in the introduction to his *Avantisundarī*. Forced to leave the Tami<u>l</u> South during a time of great famine and chaos, he is eventually invited to return to the Pallava court (perhaps by Narasiṁhavarman II, to whom we shall return in a moment), where he meets great builders and architects and travels to the Pallava seashore city of Māmallapuram.[21]

Not only does Pallava-era literature in a variety of languages and from a variety of religious perspectives dwell at length on the rains and the specter of famine, but Pallava monuments themselves also bespeak a concerted interest in water imagery, manipulation, and management. This is certainly evident at the site just mentioned in connection with Daṇḍin's visit: Māmallapuram, the unfinished, oft-studied, but still mysterious Pallava seaside city built largely by Narasiṁhavarman I in the mid-seventh century and named in honor of one of his royal epithets, Mahāmalla, or "The Great Wrestler."[22] The site is a complex one, with various temples and shrines exhibiting a variety of styles, from caves and rock-cut temples to a free-standing temple on the beach (built by the later ruler, Narasiṁhavarman II). New archaeological discoveries are continually being made here, particularly since the devastating tsunami of late 2004 radically altered the coastline.[23] Of particular interest given this essay's topic is the prominence of water imagery (and presumably water-related ritual praxis) at the impressively sculpted open-air hillside stone bas-relief known as the "Great Penance Panel" or sometimes "Descent of Gaṅgā." Virtually every element of this complex scene has generated scholarly debate, from the figure of the sage standing in stoic ascetic posture (is this Arjuna, one of the five Pāṇḍava brothers, or the sage Bhagīratha, entreating the gods to send relief from a terrible drought?)[24] to the playful imagery of mimicking cat and gullible mice and all the beautiful imagery of *nāga*s or serpents, elephants, and celestial beings. Śiva also stands there expectantly by the sage, portrayed as waiting to break the fall of the powerful river Gaṅgā with his matted locks. For a variety of reasons that extend beyond the scope of this essay, I understand the Great Penance Panel to narrate the great sage Bhagīratha's penance that results in the descent of the mighty (and holy) river Gaṅgā, a visual depiction of a common

Pallava inscriptional claim that the rise of their dynasty resembled the descent of the beneficent Gaṅgā from the heavens.[25] More than one thousand miles away from the river Gaṅgā and with no river of its own near the sculpted rock, the sight, during heavy monsoon rains, of water pouring down (like the Gaṅgā herself) through the artificially enhanced cleft in the rock must have been impressive. Evidence has also been found, however, of a large cistern that allowed for an artificial "descent" to coincide, perhaps, with various kinds of ritual activities. For the centerpiece of his seaport city, in other words (and the Penance Panel does, indeed, seem to lie very much at the center of the excavated Māmallapuram monuments), Narasiṁhavarman I chose to emphasize the connections among water, Śiva, and the dynasty itself, the natural world in concert with the manipulated or managed world.

Water imagery and technology—and particularly water in the context of the ritual technologies of the temple—also predominates in the inland Pallava capital of Kāñcīpuram. Laid out at the confluence of the rivers Pālar and Vēkavati perhaps a thousand years or more before the Pallavas, the Pallavas themselves expanded the city carefully in the form of a *maṇḍala*, with a royal palace at the center and temples to Śiva, Viṣṇu, and the local form of the goddess laid out facing each other on an east-west axis, with Jain and Buddhist institutions most likely to the south and west.[26] In the iconography of the Vaiṣṇava temple attributed to Nandivarman II, the Vaikuṇṭha Perumāḷ, studied in depth by the late Dennis Hudson, the world is envisioned as "an open lotus, floating in an inconceivably huge pond of pure water contained in a placenta-like envelope at the center of God";[27] in the ritual life of the temple, which would seem to have involved both exoteric and esoteric attempts to manipulate the natural world, drainage pathways around the central images could be plugged and filled with water.[28] Indeed, in Kāñcīpuram and throughout the territory they controlled for more than three centuries, the Pallavas constructed and patronized both rock-cut and cave shrines, as well as the first free-standing stone temples and tanks, drawing a strong connection among the power of the deities enshrined, the rituals performed in their name, and the health and well-being of the region.

Pallava monarchs significantly altered the landscape of their realm, developing a series of tanks and irrigation systems to allow dry crop cultivation to be replaced by reliable wet crop cultivation.[29] The Pallavas were called *kāṭuveṭṭis*, "destroyers of forests," for their efforts to clear more agriculturally viable land—highlighting a clear interest in land management and development over any contemporary notion of wilderness preservation.[30] Over three centuries a large number of tanks, wells, and irrigation canals were constructed, many of them in continuous use down to the present day. Inscriptions record the construction and maintenance of at least thirteen major tanks or man-made reservoirs, often in conjunction with the building of a temple or shrine;[31] numerous wells were also dug, and permission to dig wells was specifically granted to Brahmin settlements (*brahmadeya*) by the king.[32] Inscriptions also record—as royally sponsored activities of significance—the construction of irrigation canals, spring-fed channels, sluices, and simple irrigation apparatus of bucket, pole, and counterweight; equally important was the establishment of local committees to manage and maintain the waterworks.[33] Water management, in other words, constituted a central concern of all Pallava rulers, both in a practical, on-the-ground sense and in the larger powers harnessed by ritual and religious technologies (the details of which unfortunately remain obscure).

At least one Pallava-era text—albeit quite tersely—draws a strong connection among the pragmatics of water management, the timely coming of the monsoon rains, and what we would today call "ethics." Recall the musings of the unnamed Kāñcīpuram king in the Buddhist *Maṇimēkalai* as he wondered why the rains had failed his kingdom. He wondered not about global climactic shift or unfavorable gods, but about the moral character of both himself and his citizenry and the integrity of their ritual practices: "Was it the straying of the royal scepter? Was it some error in performing austerities? Was there some imperfection in the virtue of the women whose hair is full of honey-laden blossoms?" The king, in other words, ties what today we might call the virtues—or lack thereof—of himself and his individual subjects to the condition of the world; he fears that some interior lack of alignment, justice, or virtue has necessarily resulted in an exoteric misalignment of the planets resulting in the failure of the rains. This connection—between individual human virtue and the creation and sustenance of an ecology of human flourishing—is made more positively in the early Pallava (perhaps sixth century) poetic text on ethics known as the *Tirukkuṟaḷ*. In this short yet poetically beautiful text, a series of verses outline and praise the virtuous practices of king and subject, ascetic and layperson. Two people in particular are charged in the *Tirukkuṟaḷ* with ensuring the timely arrival of the monsoon rains, and they make for quite a surprising pair. The first might well be expected, given the discussion above. In the midst of an extended treatment of the virtues necessary to rule wisely, the poet states: "Rains and crops appear regularly in the land of the righteous king,"[34] while "those who live with an ungracious king suffer like the earth without rain,"[35] and "if the king rules unjustly, then the skies withhold the rains."[36] The poet, in other words, envisions a strong and important connection between the moral order of the king and the order of the natural world that governs the rains; given the Pallava-era focus on water in temples, monuments, and agricultural development, this linking of royal ethics to prosperity and well-being is perhaps not surprising. What is unexpected, however, is the sole other figure granted the power to affect, even command, the rains in the *Tirukkuṟaḷ*. In the midst of a lengthy treatment of the virtues of the domestic life, the *Tirukkuṟaḷ* boldly wonders: "Is there anything greater than a woman rooted in womanly virtue?[37] Arising and revering her husband as lord, she says 'Rain,' and it rains."[38] Here again, the virtues that the poet goes on to describe in terms of fidelity, restraint, gracefulness, and compassion—the same qualities of the just king transferred to the domestic sphere—command the order of the exoteric world, the coming of the all-important rains that the Pallavas endeavored to ensure and harness. In the intricate relationship between personal moral order and world order that is as old as the hymns of the *Rig Veda* and the *Bhagavad-gītā*, the welfare of the larger world is entirely dependent on the ordering of the lives of its individual components.

Rājasiṁha's Three Temples

With all of the above in mind—both the general activities of the Pallava rulers and this specific tying of individual moral order to the wider ecology of human flourishing envisioned by them—let us take a brief look at the specific projects of one of the most productive builders and generous patrons among the Pallavas, Narasiṁhavarman II, more commonly known by his epithet, Rājasiṁha or "Lion among Kings," who is generally believed to have ruled from about 690 to 729.

This relatively long and stable rule emerges from the inscriptional record as comparatively free of military engagement with the Pallavas' archrivals to the north and west, the Cālukyas. His reign may also have begun at the end of a long and disastrous famine in the Pallava realm.[39] In one temple inscription at Panamalai, of Rājasiṁha's rule it is said: "the shade of *dharma* flourished, even though scorched by the cruel heat of the Kali *yuga* [the cosmic age of moral blackness]."[40] Some scholars have speculated that Rājasiṁha was a contemporary of the great Śaiva poet-saint cited above, Cuntarar;[41] others have argued that one of his titles, *paramaheśvara* or "Supreme Maheśvara," indicates that he received formal initiation (*dīkṣā*) into the branch of Śaiva thought and practice known in his day as the Śaiva Siddhānta.[42] Rājasiṁha was, indeed, a Śaiva, and also a great patron of the courtly arts; as above, it is believed that the poet and literary theorist Daṇḍin was attached to his court.[43] Rājasiṁha also sponsored a great many innovative building projects, from irrigation canals to temples. The latter—temples—will be briefly considered here for what they might suggest regarding the king's vision of human flourishing. Rājasiṁha's temple projects mark a quite radical departure from those of his predecessors; under his patronage, the earlier focus on carving shrines from caves and natural rock gives way to the construction of freestanding structures made of stone, brick, and plaster. All of his temples are dedicated to Śiva and all share certain unique qualities that are, one might argue, relevant to the theme of ecology and human flourishing.[44] The three temple structures that are indisputably attributed to Rājasiṁha are the Shore Temple at Māmallapuram, the Kailāsanātha Temple in Kāñcīpuram, and the Tālapurīśvara Temple at Panamalai in the Villupuram District. All three bear quite obvious architectural, epigraphic, and iconographic similarities that mark genuine departures from earlier styles.

First, all three temples face eastward, in the direction of the Vedic king of the gods, Indra, ruling over the deities of heaven.[45] Each central shrine (*garbhagṛha*, literally "womb house") contains a distinctively Pallava *liṅga* (aniconic "mark" of Śiva) that would become standard for most Tamiḻ Śaiva shrines down to the present day. The goddess figures prominently in each of the three temples, in forms both gently domesticated and martially ferocious. Indeed, Rājasiṁha's temples iconographically introduce Śiva as "family man" for the first time in the Tamil̲-speaking South: the inner wall of the *garbhagṛha*, behind the *liṅga* in each temple, presents Śiva in his Somāskanda form, literally the lord "with (*sa*) Umā (his consort) and Skanda (his second son)."[46] Rājasiṁha's temples also introduce to Tamil̲ temple iconography for the first time Śiva's more pan-Indically celebrated elder son, the elephant-headed Gaṇeśa. The juxtaposition, even blending, of male and female divine forms is also manifest in Rājasiṁha's interest in Śiva as Ardhanārīśvara, the lord who is literally half man and half woman, at Kailāsanātha[47]—as at Māmallapuram[48]—depicted holding the musical *vīṇā* (a stringed instrument) as the purveyor of the cultural arts. One art historian, Padma Kaimal, suggests reading the entire Kailāsanātha Temple—hinting that others might be read this way as well—as a union of male and female aspects of divinity, with a complex series of goddess images surrounding the central, and centrally male, *liṅga*.[49] In all three temples, such divine iconography is, in another Pallava innovation in South India, swathed in beautiful depictions of the natural world.[50]

Such imagery of the divine—both male and female—in its many forms is organized in Rājasiṁha's temples around the king himself, particularly in the Kailāsanātha Temple in his capital city of Kāñcīpuram, where 252 royal titles (*biruda*) are inscribed on the façades of the smaller shrines surrounding the *garbhagṛha* in four distinctly southern scripts (although the language of all is Sanskrit). The titles themselves are quite fascinating, ranging from "foremost among kings" (*ekarāja*) and "the subduer of rebels" (*udvṛttadamana*) to "foremost among the handsome" (*ekasundara*), "refuge of the distressed" (*taptaśaraṇa*), and "[bringer] of continual rains" (*nityavarṣa*).[51] We will return to these honorifics in a moment, but, simply in terms of visual presentation, the glory of the deity and the power and grace of the king are intimately bound together.

Another striking innovation marked by Rājasiṁha's temples at Māmallapuram, Kāñcīpuram, and Panamalai is the very architecture itself, departing from the open-air, public atmosphere of earlier Pallava structures to enclose both central image and the entire site behind high walls. While the mountain-like *vimāna* housing the central image would be visible from afar, the sacred images themselves—as well as the human ritual activity of the temple—would be entirely hidden from public view for the first time. The worshipper is entirely enclosed and, at times, as he approaches the *liṅga*, plunged into total darkness. The enclosing walls also allow for smaller, subsidiary shrines to be constructed surrounding the *garbhagṛha*, as at Kailāsanātha, where one such shrine is attributed to Rājasiṁha's son and eventual successor, Mahendravarman III.[52]

Rājasiṁha's Vision of Human Flourishing

What, if anything, might all of these innovations on the part of Rājasiṁha suggest about his vision of human flourishing, of the "ecology" that enables such human flourishing as he understood it? To begin with his iconographic creativity, it is tempting to see his marked preference for highlighting Śiva as Somāskanda right behind the central *liṅga* and on temple panels as an integrative or mediating move, incorporating into southern Śaiva iconography—as Śiva's younger son, Skanda—the extremely important Tami<u>l</u> deity Muruka<u>n</u>, lord of the hills and lover to Vaḷḷi, an unlettered local girl of the mountains.[53] Rājasiṁha's strong preference for including imagery of the feminine divine—as the maternal Umā in the Somāskanda panels, as the half of Śiva who bears human cultural forms in Ardhanārīśvara with the *vīṇā*, and in the wide variety of demure and demonic female forms facing the central shrine at Kailāsanātha—also seems to speak to an integrative or mediating project, calling on *all* forms of the divine, domestic and ascetic, male and female. These multiple forms, as at Kailāsanātha, are also portrayed quite explicitly as organizing and managing the natural world.

Yet Śaiva divinity, in its myriad manifestations at Rājasiṁha's three temples, manages the natural world—controlling the rains, bringing peace and prosperity—seemingly only in concert with the king himself, as the royal *biruda* inscribed on each temple, but particularly at Kailāsanātha, make clear. The celebration of Rājasiṁha's 252 great attributes—including, quite ironically, *guṇavinīta*, "he whose virtue is modesty"!—ranges from praising his heroism in battle (as in *paracakramarddana*, "the destroyer of hostile enemies," and *āhavadhīra*, "steady in battle") to his patronage of culture (as in *kāviprabodha*, "he of poetic insight"), his wisdom (as in *tatvavedī*, "philosopher"), and his virtue (as in *dharmmanitya*, "he who ever abides in *dharma*"). Of particular interest in this context are the numerous references to Rājasiṁha's ability to bring the rains on time and, in so doing, he is also often likened to the monsoon cloud itself: he "appears like a rain-cloud" (*parjjanyarūpa*) and is "the cloud that showers prosperity" (*śrīmegha*).[54] Facing, as these inscriptions do on four sides, Śiva as both *liṅga* and Somāskanda in the central shrine (*garbhagṛha*), the visual implications are clear: the king's ability to rule wisely and the manifold forms of the divine are mutually interdependent in their shared project of sustaining an ordered world of balance and plenty.

How this interdependence was understood to manifest and do its proper work in the world is now something of a matter for historical speculation, but a clue perhaps lies in both the architecture of the Rājasiṁha temples and the repeated inscriptional assertion that Rājasiṁha was himself a Śaiva ritual initiate, a "follower of the Āgamas" (*āgamānusāri*), "one whose authority is the Āgama" (*āgamapramāṇa*), one who is "devoted to tradition" (*ācārapara*).[55] While a full discussion of what Āgama means in this context lies beyond the scope of this paper, the Āgamas are, in brief, post-Vedic texts that outline a series of ritual technologies for both self-transformation and the generation of powers for the glory of Śiva and the good of the world.[56] Such practices require increasingly complex rituals of initiation (*dīkṣā*) under the guidance of a teacher, and move from the more exoteric to the increasingly esoteric. Eighty years ago, Longhurst speculated that Rājasiṁha's architectural innovations—enclosing the temple's images and the practices of the worshipper and cutting both off from public view—heralded a new religious focus on "secrecy and mysticism."[57] More recently, in the Vaiṣṇava context of the Vaikuṇṭha Perumāḷ Temple constructed some fifty years after the Kailāsanātha in Kāñcīpuram, Dennis Hudson has argued at length for a close correlation among temple architecture, the equivalent of Vaiṣṇava Āgamic ritual practices, and specific Vaiṣṇava narrative literature, all aimed at both the self-transformation of the king himself and the maintenance of the well-being of his kingdom.[58] Rājasiṁha, in other words, a self-proclaimed "follower of the Āgamas," constructed a royal temple at Kāñcīpuram to perform the esoteric rituals of one Śaiva path, thus generating the virtues, qualities, and capabilities loudly proclaimed in the temple *biruda*. Such ritual practice—encompassing knowledge (*jñāna*), ritual action (*kriyā*), proper conduct (*caryā*), and physical and mental discipline (*yoga*)—formed both the royal practitioner's inner world, character, and trajectory (ultimately toward liberation) and at the same time yielded, via divine grace, an orderly and prosperous kingdom.[59]

In this context, Pallava patronage of multiple religious institutions and ritual practices in a number of languages seems less a move toward ideological tolerance or inclusiveness and more a pragmatic effort to tap into all ritual powers of transformation in order to secure an ordered world, perennially measured in terms of the timely arrival of the monsoon rains. All Pallava rulers, as noted above, appear to have patronized all aspects of the diverse religious landscape of Tamiḻ-speaking South India, including not only Śaivas and Vaiṣṇavas, but Jains and Buddhists as well. The evidence that these Pallava "forest-killers" have left behind would appear to indicate that, like patronage of multiple religious communities and their ritual technologies, religious and literary imagery of the natural world likewise need not imply any love of nature or appreciation for the wilderness. Rather, Pallava literature, iconography, and building projects focused on materially and ritually *managing* the natural world via all possible means to avoid the ever-looming crises of drought and famine.

In this context, the somewhat surprising location of Rājasiṁha's third temple to Śiva at Panamalai begins to make some sense. That Rājasiṁha would construct royal temples in both his capital city (Kāñcīpuram) and his lineage's seaside port city (Māmallapuram) requires no detailed explanation. But why would he build a temple on a rocky outcrop hundreds of miles to the south and west, in Panamalai?

The Panamalai temple bears much resemblance to Rājasiṁha's other buildings, from the preference to Somāskanda to the general blueprint of the complex. A single inscription can safely be attributed to Rājasiṃha, outlining his dynastic lineage and naming the king himself as a mighty "vanquisher of elephants."[60] Yet why here? The beginning of an answer might be tentatively found in both the reference to the king's military valor and the temple's location next to a large lake in a district traditionally well watered by tributaries of the Kāveri and Pālar rivers.[61] Here references to military prowess at the side of a lake marking a Pallava southern frontier dominated by rich water sources surely invites one to speculate that land and water management are somehow thoroughly implicated in Rājasiṁha's palace for Śiva at this particular spot.

Moral Character and Human Flourishing

What emerges from the extant but fragmentary literary, material, and inscriptional record of the Pallavas provides but a glimpse into a vision of human flourishing founded on both individual and royal moral development, on moral "character" if you will, and complex ritual technologies intended to manage, in concert with the divine, both one's own self-transformation and the transformation of the natural world into a place of order, predictability, and prosperity. One cannot be certain if this vision of human flourishing worked and, if so, for whom. By the end of the ninth century, the Pallavas were overtaken by their enemies to the South, the Cōḻas, who would build on Pallava projects but take them in new directions. What constituted human flourishing in the Pallava context—at least from the elite level of its extant remnants—obviously stands at some remove from contemporary ideals of democracy and self-determination: Pallava order—both moral and natural—depended on the ritual life of a just king working in concert with a compassionate if at times unpredictable set of divine beings. Yet the ecology of human flourishing that Pallava materials present, perhaps taken at its most basic level, is somewhat less foreign to us today, with its sharp connections drawn between personal morality and the well-being of a fragile world continually susceptible to drought, famine, and decay. Perhaps what sets the Pallava vision of human flourishing both individually and collectively apart from our own is its seeming emphasis on ritual praxis as the primary means of connecting person to world, divinity to humanity, individual virtue to collective well-being.[62] What, if anything, might we do with that insight? That is perhaps a question best answered by others, but for the historian of religion it certainly points the way toward interpreting ritual praxis in the broader context of ethics, theology, literature, and agriculture.

Glossary of Selected Words

Term (English)	***Explanation***
Āgamas	post-Vedic texts offering instructions for transformation (self and otherwise) through a variety of ritual and devotional practices
Ardhanārīśvara	the Hindu god Śiva, in half-man, half-woman form
Arjuna	one of the Pāṇḍava brothers, a hero of the Mahābhārata, favored by Kṛṣṇa
Bhagavad-gītā	a Hindu scripture spoken by Kṛṣṇa to Arjuna in the Mahābhārata
Bhagīratha	great king or sage in Hindu mythology, bringer of the Ganges to earth
bhakta	one who lives a life of devotion, i.e., a saint
bhakti	means "devotion" and refers to an intense personal and emotional relationship with the Divine
Brahmin	Hindu priestly and scholarly class
Caṅkam (Sangam)	classic Tamil̲ literature from before the fourth century
Cen̲n̲ai (Chennai)	city of more than six million people in northeast Tamil Nadu, formerly known as Madras
Cōl̲a (Chola)	dynasty ruling in Tamil Nadu from the ninth to twelfth centuries, also mentioned in Caṅkam literature as a pre-Pallava dynasty
dharma	the underlying principles of cosmic and social order, as well as one's obligation to act in accordance with those principles
dīkṣā	formal initiation
Gaṇeśa (Ganesh or Ganesha)	elephant-headed god, son of Śiva and Pārvatī
Gaṅgā (Ganges)	the most sacred river of India, which flows more than one thousand miles from the Himalayas out to the Bay of Bengal; also the name of the goddess who embodies that river
garbhagṛha	a temple's central shrine (literally, a "womb house")

Jain	relating to Jainism, an Indian religion stressing nonviolence
Jina	one who has reached enlightenment in Jainism
Kailāsanātha Temple	temple in Kāñcīpuram, built by Narasiṁhavarman II
Kāñcīpuram (Kanchipuram)	a holy city in South Indian Hinduism, in the northeast corner of the Indian state of Tamil Nadu (about seventy kilometers southwest of Ceṉṉai), capital of Pallava dynasty
Kṛṣṇa (Krishna)	a human-divine incarnation of Viṣṇu
liṅga (lingam)	a phallic shape associated with Śiva as his aniconic symbol
Mahābhārata	ancient Indian epic
Mahāmalla	epithet for Narasiṁhavarman I, meaning "Great Wrestler"
Māmallapuram (Mamallapuram)	an archaeological site and small town in northeast Tamil Nadu, about sixty kilometers south of Ceṉṉai, home to many monuments, including the "Great Penance Panel"
Maṇimēkalai	Tamiḻ Buddhist text from the sixth or seventh century featuring a woman who becomes a nun
Murukaṉ	local Tamiḻ deity integrated into pan-Indian Hinduism as Śiva's second son, Skanda
Nandivarman II	eighth-century Pallava ruler, builder of Vaikuṇṭha Perumāḷ Temple in Kāñcīpuram
Narasiṁhavarman I	seventh-century Pallava ruler, builder of Māmallapuram; see also Mahāmalla
Narasiṁhavarman II	late–seventh- and early–eighth-century Pallava ruler, also known as Rājasiṁha
Pallava	dynasty ruling in northern Tamil Nadu from the seventh to ninth centuries, also mentioned in earlier Caṅkam literature
Pāṇḍava	the five brothers who are heroes of the Indian epic, the *Mahābhārata*
Pārvatī	Śiva's consort, also known as Umā

Rājasiṁha	epithet for Narasiṁhavarman II, meaning "Lion among Kings"
Rig Veda	collection of hymns, part of the canon of Hinduism, one of the earliest texts known in the world
Śaiva (Shaiva)	devotee of Śiva, relating to Śiva
Śaiva Siddhānta	particular philosophical and ritual movement devoted to Śiva
Sanskrit	classical Indian language in which many sacred texts are written
Śiva	one of the primary supreme divinities in the Hindu pantheon
Skanda	Śiva's second son
Somāskanda	Śiva as a family man, with his consort and second son Skanda
Tālapurīśvara Temple	stone temple, built by Rājasiṁha in Villupuram District
Tamiḻ (Tamil)	Dravidian language spoken in South India, Sri Lanka, and Southeast Asia; also the related culture and people
Tamil Nadu	contemporary state in southeastern India, capital Ceṉṉai
Umā	Śiva's consort, also known as Pārvatī
Vaikuṇṭha Perumāḷ	temple in Kāñcīpuram, dedicated to Viṣṇu
Vaiṣṇava (Vaishnava)	devotee of Viṣṇu
Veda	ancient Sanskrit scriptures of India, for example *Rig Veda*
Vedic	relating to the Vedas or of the time period when the Vedas were composed, about 1500–500 BCE
vimāna	tower over the innermost shrine of a Hindu temple
vīṇā	stringed instrument, somewhat similar to a lute
Viṣṇu (Vishnu)	one of the primary supreme divinities in the Hindu pantheon

Notes

1. Editors' note: We offer this note to orient general readers who may not be familiar with the traditions and history of South Asia. This paper discusses the cultural and literary traditions of northern Tamil Nadu. Tamil Nadu is about the size of Greece or the land mass of New York State; it lies in the southeastern corner of India, bordered to the east by the Indian Ocean. The largest city of Tamil Nadu, with more than six million people, is Ceṉṉai (formerly known by its colonial name of Madras). At the end of the essay, we have created a brief glossary of words, some of which may be familiar to the reader in their English renditions and not in the Sanskrit, such as Shiva rather than Śiva. For an overview of Hinduism, the religion at the heart of this paper, see Gavin Flood, *An Introduction to Hinduism* (Cambridge: Cambridge University Press, 1996). For a general overview of Tamiḻ literature, see Kamil V. Zvelebil, *The Smile of Murugan: On Tamil Literature of South India* (Leiden: E. J. Brill, 1973). For more on Hinduism and ecology, see the book by that name, edited by Christopher Key Chapple and Mary Evelyn Tucker (Cambridge, MA: Center for the Study of World Religions, 2000). Other books in the series related to other religions mentioned in this paper might also be helpful, all published by the CSWR: Mary Evelyn Tucker and Duncan Ryûken Williams, eds., *Buddhism and Ecology: The Interconnection of Dharma and Deeds* (1997); and Christopher Key Chapple, *Jainism and Ecology: Nonviolence in the Web of Life* (2002).
2. "Ecologies of Human Flourishing: 2009–2010 Lecture Series at the Center for the Study of World Religions, Harvard Divinity School," brochure, overleaf.
3. For the general dating of the earliest Tamiḻ inscriptions in Brāhmī script, all seemingly Jain, see Iravatham Mahadevan, *Early Tamil Epigraphy: From the Earliest Times to the Sixth Century A.D.*, Harvard Oriental Series, ed. Michael Witzel, vol. 62 (Chennai: Cre-A; and Cambridge: Harvard University Press, 2003).
4. Burton Stein, *Peasant State and Society in Medieval South India* (Delhi: Oxford University Press, 1985), 26.
5. Cadambi Minakshi, *Administration and Social Life under the Pallavas* (1938, Madras: University of Madras, 1977), 1–3, briefly summarizes the types of sources available for examining Pallava history, culture, and society.
6. J. Stephen Lansing, *Perfect Order: Recognizing Complexity in Bali* (Princeton: Princeton University Press, 2006).
7. See Minakshi, *Administration and Social Life*, 3–8, for a summation of the principal theories.
8. The *Perumpāṇāṟṟuppaṭai*, or "A Long Poem for Bards with Lutes," attributed to Kaṭiyalūr Uruttiraṅkaṇṇaṉār, praising a second-century ruler of Kāñcīpuram named Toṇṭaimāṉ Iḷantiraiyaṉ. See Kamil V. Zvelebil, *A Lexicon of Tamil Literature* (New York: E. J. Brill, 1995), 549.
9. See Michael D. Rabe, *The Great Penance at Māmallapuram: Deciphering a Visual Text* (Chennai: Institute of Asian Studies, 2001), 159–160, for the current scholarly consensus on primary names and order of succession.
10. D. Dennis Hudson, *Krishna's Mandala: Bhagavata Religion and Beyond*, ed. John Stratton Hawley (New Delhi: Oxford University Press, 2010), 49.

11. For an accessible introduction to Caṅkam poetics, see A. K. Ramanujan; *Poems of Love and War: From the Eight Anthologies and the Ten Long Poems of Classical Tamil* (New York: Columbia University Press, 1985), 229–297.
12. For a recent study of the tension between *nāṭu* (cultivated country) and *kāṭu* (wilderness) among the Kallar communities of Tami̲l-speaking South India both today and in history, see Anand Pandian, *Crooked Stalks: Cultivating Virtue in South India* (Durham: Duke University Press, 2009).
13. Appar 6.95.5: *tirukkōyil illāta tiru il ūrum / . . . aṭavi—kāṭē*. All hymns of the Śaiva saints are drawn from the critical edition: *Tēvāram: Hymnes Śivaites du Pays Tamoul*, ed. T. V. Gopal Iyer, 3 vols. (Pondichery: Institut Français d'Indologie, 1984–1991). All translations, unless otherwise noted, are the author's own.
14. Appar 6.87.1: *vaṭamoḻiyum tem̲tamiḻum maṟaikaḷ nāṉkum / āṉavaṉ kāṇ . . . / civaṉ avaṉ kāṇ—civapurattu em celvaṉ tāṉē.*
15. See Ramanujan, *Poems of Love and War*, 242.
16. Cāttaṉār, *Maṇimēkalai*, ed. U. Vē. Cāminātaiyar (Ceṉṉai: Ṭākṭar U. Vē. Cāminātaiyar Nūlnilaiyam, 1981), 28.188–191: *ceṅkōl kōṭiyō ceytavam piḻaittō / koṅkaviḻ kuḻalār kaṟpukkuṟai paṭṭō / nalattakai nallāy naṉṉāṭu ellām / alattaṟkālai ākiya taṟiyēṉ.*
17. Appar 6.98.1: *nām ārkkum kuṭi allōm namaṉai añcōm / . . . . piṇi aṟiyōm.*
18. Cuntarar 7.20.1: *vāṭi varuntōm / . . . kuṇṭaiyūr cila nellu peṟṟēṉ.*
19. Cuntarar 7.34.1: *immaiyē tarum cōṟum kūṟaiyum.*
20. Campantar 1.92.1: *kācu nalkuvīr . . . miḻalaiyīr.*
21. Daṇḍin, *Avantisundarī*, Trivandrum Sanskrit Series, vol. 172, ed. K. S. Mahādeva Śāstri and Nārāyaṇa Piḷḷai (Trivandrum: S. K. Pillai, 1954). A. K. Warder discusses both the content of the narrative and Daṇḍin's seemingly autobiographical introduction in his *Indian Kāvya Literature* (Delhi: Motilal Banarsidass, 2009), vol. 1, 211–212, and vol. 4, 166–209.
22. Scholars have debated both the dating and the patronage of the monuments at Māmallapuram for more than a century, although a general consensus has emerged attributing the bulk of the building to Narasiṁhavarman I. For a recent discussion of the various arguments, see Rabe, *Great Penance at Māmallapuram*, 50–58.
23. See, for example, Paddy Maguire, "Tsunami Reveals Ancient Temple Sites," *BBC News Online* at http://news.bbc.co.uk/2/hi/south_asia/4312024.stm.
24. Rabe concludes in the *Great Penance* cited above that *both* stories are deliberately intended.
25. For a compelling argument for this point of view, see Michael Lockwood, *Māmallapuram and the Pallavas* (Madras: Christian Literature Society, 1982), 6–12.
26. For a discussion of Kāñcīpuram's *maṇḍala* layout, see D. Dennis Hudson, "Ruling in the Gaze of God: Thoughts on Kanchipuram's Maṇḍala," in *Tamil Geographies: Cultural Constructions of Space and Place in South India*, ed. Martha Ann Selby and Indira Viswanathan Peterson (Albany: State University of New York Press, 2008), 87–116.
27. D. Dennis Hudson, *The Body of God: An Emperor's Palace for Krishna in Eighth-Century Kanchipuram* (New York: Oxford University Press, 2008), 304.
28. Ibid., 17.

29. Stein, *Peasant State and Society*, 68.
30. Minakshi, *Administration and Social Life*, 183.
31. Ibid., 118–125.
32. Ibid., 126.
33. Ibid., 126–134.
34. *Tirukkuṟaḷ* 545: *iyalpuḷi kōlōccu maṉṉavaṉ nāṭṭa / peyalum viḷaiyuḷum tokku.* All *Tirukkuṟaḷ* quotations are drawn from *Tirukkuṟaḷ mūlamum Parimēlaḻakar uraiyum* (Ceṉṉai: Teṉṉintiya Caivacittānta Nūṟpatippu Kaḻakam, 1991).
35. *Tirukkuṟaḷ* 557: *tuḷiyiṉmai ñāllatiṟkku eṟṟaṟṟē vēntaṉ / aḷiyiṉmai vāḻum uyirkku.*
36. *Tirukkuṟaḷ* 559: *muṟaikōṭi maṉṉavaṉ ceyyiṉ uṟaikōṭi / ollātu vāṉam peyal.*
37. *Kaṟpu*, often unhelpfully rendered into English as "chastity," seems to have greater resonance with the Sanskrit term *śakti*; *kaṟpu* is the tremendous power afforded women who are able to control their sexuality and wield that power wisely.
38. *Tirukkuṟaḷ* 54–55: *peṇṇiṉ peruntakka yā uḷa kaṟpu eṉṉum / tiṇmai uṇṭāka peṟiṉ / teyvam toḻā aḷ koḻunaṉ toḻutu eḻuvāḷ / peyya eṉa peyyum maḻai.*
39. No Pallava records note this famine, but both Chinese and later South Indian sources are suggestive. See T. V. Mahalingam, *Kāñcīpuram in Early South Indian History* (Bombay: Asia Publishing House, 1969), 109–110.
40. T. V. Mahalingam, *Inscriptions of the Pallavas* (New Delhi: Indian Council of Historical Research, 1988), 201: *chāyāsamuddhāma . . . jayati kaliyugagrīṣmataptopi dharmaḥ.*
41. See Indira Viswanathan Peterson, *Poems to Śiva* (Princeton: Princeton University Press, 1989), 19, n. 1.
42. Hudson, *Body of God*, 57.
43. See Warder's discussion of Daṇḍin's *Avantisundarī* in *Indian Kāvya Literature* cited above, as well as Robert DeCaroli, "An Analysis of Daṇḍin's *Daśakumāracarita* and Its Implications for Both the Vākāṭaka and Pallava Courts," *Journal of the American Oriental Society* 115, no. 4 (1995): 671–678.
44. A. H. Longhurst, *Pallava Architecture, Part III: The Later or Rājasiṁha Period* (New Delhi: Cosmo Publications, 1982), 1.
45. See Hudson, *Body of God*, 106–109, for a general discussion of the cardinal directions and their symbolic meanings in the context of the Pallava-era Vaikuṇṭha Perumāḷ Vaiṣṇava Temple in Kāñcīpuram.
46. For a discussion of the iconography and meaning of Śiva as Somāskanda, see H. Daniel Smith and M. Narasimhachary, *Handbook of Hindu Gods, Goddesses, and Saints* (Delhi: Sundeep Prakashan, 1997), 38–48.
47. P. V. Jagadisa Aiyar, *South Indian Shrines* (New Delhi: Rupa, 2000), 72, notes the existence of Śiva in this form at Kailāsanātha, with the female aspect bearing the *vīṇā* and the male aspect seated atop Śiva's divine bull-vehicle, Nandi.
48. See Michael Lockwood, "Vīṇādhara Ardhanārīśvara," in *Pallava Art*, ed. Michael Lockwood, A. Vishnu Bhat, et al. (Madras: Tambaram Research Associates, 2001), 234–238.
49. Padma Kaimal, "Learning to See the Goddess Once Again: Male and Female in Balance at the Kailāsanāth Temple in Kāñcīpuram," *Journal of the American Academy of Religion* 73, no. 1 (2005): 45–87.

50. Ibid., 71–74, describes female deities surrounded by "elephants, water buffalo, rams or antelope . . . suggest[ing] a wilderness, a place profoundly unlike the urban landscape the Kailāsanāth temple has occupied."
51. See Michael Lockwood, "Royal Titles of Rājasiṁha and Mahāmalla," in *Pallava Art*, 173–188.
52. Longhurst, *Pallava Architecture, Part III,* 13.
53. For a detailed study of the trajectory of Murukaṉ in Tamiḻ cultural history, from his appearance in Caṅkam literature to his enthronement as the official deity of the Indian state of Tamil Nadu in 1970, see Fred W. Clothey, *The Many Faces of Murukaṉ: The History and Meaning of a South Indian God* (New York: Mouton, 1978).
54. Drawn from Lockwood, "Royal Titles," 179–185.
55. Ibid., 180.
56. For an excellent study of Āgamic ritual practice, see Richard H. Davis, *Ritual in an Oscillating Universe: Worshiping Śiva in Medieval India* (Princeton: Princeton University Press, 1991).
57. Longhurst, *Pallava Architecture, Part III,* 15.
58. Hudson, *Body of God.*
59. Davis, *Ritual in an Oscillating Universe,* 10.
60. Longhurst, *Pallava Architecture, Part III,* 7.
61. See the list of rivers and basins for Villupuram District at http://www.aedatlas.tn.nic.in/tnwshed.php?dt_code=07. A discussion of the environmental pressures on various water sources in the Villupuram District can be found at http://muthupages.blogspot.com/2007/02/history-of-water-bodies-in-villupuram.html.
62. See Adam B. Seligman, Robert P. Weller, and Michael Puett, *Ritual and Its Consequences: An Essay on the Limits of Sincerity* (New York: Oxford University Press, 2008), for an extended discussion of the ways in which ritual practices shape practitioners to live in a fragile world constantly tending toward imperfection and decay.

Love and Longing in the Time of Rain

A Response to Monius's *Ecologies of Human Flourishing*

Archana Venkatesan

It is particularly appropriate that an essay on the ecology of human flourishing in medieval South India should focus on the enigmatic Pallava kings, whose obscure and much debated dynastic name suggests a leaf or branch (*pallav*).[1] The Pallavas were path-breakers in many regards, and their complex vision of human flourishing presents us with a number of challenging opportunities to unpack notions of landscape, spatial geography, temple building, and ritual activity in early-medieval South India. Keeping this in mind, I am going to venture out further on the limb on which Anne Monius started her essay, with its primary focus on the Śaiva literary traditions under the Pallavas. In my response, I focus on three historical-literary moments—the Caṅkam period (just before the Pallavas), Tamiḻ Vaiṣṇava *bhakti* (at the same time as the Pallavas, but in a different cultural tradition), and in conclusion the contemporary period. How do poets in each of these literary periods use landscape, and particularly the motif of the evocative rain-bearing monsoon clouds? How do notions of morality and virtue connect to the natural environment and the rain, which sustains it? How do these ideas live outside the context of poetry, altered and refigured to make sense in a contemporary, postcolonial context?

As I began to think about the ecology of human flourishing in the context of premodern South India, a famous poem from a Caṅkam anthology immediately came to mind. This poem by Tēvakkulattār, from the anthology *Kuṟuntokai,* features a nameless, faceless heroine who speaks not just of her lover, but also of the vastness of her love:

Nilattiṉum peritē vāṉiṉum uyartaṉṟu
Nīriṉum āraḷavu iṉṟecāral
Kār kōḷ kuṟiñci-p- pukkoṇṭu
Peruntēṉ iḻaikkum nāṭaṉōṭu naṭpē

Bigger than the earth, even higher than the sky
more unfathomable than the waters
is the love for the man from the mountainous land
where the *kuṟiñci* flowers with their dark stalks
yield rich honey.[2]

Kuṟuntokai 3

Much can be said about this lovely little gem of a poem: for instance, the manner in which the vastness of the landscapes telescopes into the specificity of the heroine's experience. Yet, she does not possess this man—in the verse she does not say, love for *my* man, rather it is the land (*nāṭu*) that identifies him. This word, *nāṭaṉ*, used to identify the hero signals that he is one who belongs to a *nāṭu*, a place, a country. But despite the evocation of a mountainscape (*kuṟiñci*), which according to Caṅkam poetics suggests illicit love and wildness, the *nāṭu* is a civilized place, a discrete political and social entity. Love with this civilized man, whose civility is marked by his attachment and association to the land (*nāṭu*), is on the other hand unbound, immeasurable, and by extension, uncivilized.

But how else does the cultivated, agrarian land figure in the Caṅkam literary corpus? The landscape known as *marutam*, named after the *marutam* flower, is meant to signal domesticity, where the heroine and hero after endless cycles of separation and union are finally united. It is therefore not surprising that the control of the wild, passionate desire of the *kuṟiñci* finds its parallel in the taming of the land, so that it leads to a productive end. However, it is not so simple, for *marutam* is also the site of the hero's endless infidelity, where he dashes off to consort with wily courtesans and other women. In this verse from a late anthology, *Aiṅkuṟuṉūṟu*, the heroine describes her lover "playing" with other women and then says:

Our man of the old cane town
plays partner in their love play:
        he belongs to our town,
        yet he does not.[3]

Aiṅkuṟunūṟu 15

If in the *kuṟiñci* verse the man's attachment to his land is contiguous with the love the couple shares, the violation of love, as in the *marutam* verse above, is to disown the land—he belongs to the town, but he also does not. He belongs to the woman, but he also does not. Thus, in these Caṅkam poems, the cultivation of the land does not necessarily correlate directly to the nurturing of virtue or morality for the Caṅkam poets and theoreticians. Yet, this relationship between the loving control of land to productivity and the control of the self become guiding principles for generations of Tamiḻs who follow. Indeed, I would suggest that the Caṅkam literary corpus already begins to signal the importance of the controlled and built environment to the cultivation of virtue even while exhibiting a distinctive unease with its implications, so clearly evident in the poems situated in *marutam* landscapes—simultaneously signaling female desire domesticated in the marriage and the cultivated land, while male desire thrives in the untamed natural world with its rushing rivers and wild mountain flowers.

In distinguishing the Caṅkam poetics from what the later poets under the Pallavas accomplish, Professor Monius suggests that the poetic manipulation of landscapes in the Caṅkam love poems occurs with climactic and emotional regularity. That is, each landscape always gestures toward a particular kind of meaning. *Kuṟiñci*, the landscape of the mountains, always describes secret love, while patient waiting (if indeed waiting for a lover can ever be patient!) unfolds as the rain clouds gather on the horizon and the *mullai*-jasmine begins to bloom. Though this is the general rule, Caṅkam poets do find ways to deliberately violate this emotional regularity, nowhere more clearly in evidence than when two landscapes merged, in what the Tamiḻ grammars evocatively call *tiṇai mayakkam* (the bewilderment of landscapes). *Tiṇai mayakkam* reveals the poetic limits of a landscape corralled to index human emotions and experience: the mountains are for secret love, the pastoral scape for infidelity, the wasteland for elopement and estrangement. But the natural environment is as fickle, as unpredictable, and as mercurial as love and desire, despite one's best attempts to cultivate them carefully in social and poetic contexts.

It is this subversive note, so subdued in the Caṅkam poems, that dominates the literary production of the religious communities of the Pallava period, finding sustained elaboration in the anxieties over the monsoon rains, primarily expressed in the desire for their timely arrival. The failure of the rain has a real human dimension—it brings drought, hunger, and the imminent collapse of civilization. Monius offers us the example of the sixth-century Tami<u>l</u> Buddhist epic, *Maṇimēkalai*, where the heroine wanders into the great city of Kāñcīpuram, capital of the Pallava kings, to find it brought low by a terrible drought. In the *Maṇimēkalai*, as Monius explains, the rain is intimately connected to the virtue, valor, and moral character of the king, and by extension his citizens, where drought (the failure of the rains) is a ringing indictment of the king's failed character. Of course, such links are not limited to Buddhist and Jain texts, but are prominent concerns for the devotional poet saints, particularly the Śaiva poets, who flourished under Pallava rule. For the Tami<u>l</u> Śaiva poets—Appar, Campantar, Cuntarar—Śiva devastates the natural world and consequently the human body, even while providing the cure.

What of the Vaiṣṇava Tami<u>l</u> poets (known as the *ā<u>l</u>vār*), whose domain was much further south? How did they conceptualize the relationship between rain, water, virtue, and deity? I will begin with the earliest of the three *ā<u>l</u>vār* poets—Poykai, Pūtam, and Pēy, who lived in the seventh century in the very heart of Pallava country (and who exalted the god Viṣṇu). The hagiographies of these poets tell us that they were born on three successive days from three different flowers. Poykai, whose name means pond, manifested from a golden lotus, Pūtam (ghoul) from a *mātavi* flower, and Pēy (ghost) from a red water lily. Legends aside, the birth of these three poets, who represent the genesis of Vaiṣṇava ecstatic devotion in Tami<u>l</u> country, is intimately connected to the blossoming of the natural world. This metaphor is further elaborated in the legendary account of their meeting on a stormy night. Poykai, Pēy, and Pūtam each enter the temple hall to seek shelter from the whipping winds and unrelenting monsoon rain. As the storm grows violent, and the world swells with darkness, the three poets huddle together, until they increasingly feel crowded for an unfathomable reason. Finally, they realize that their beloved god, Viṣṇu, has joined them as well. Viṣṇu, often described as the one dark as the rain clouds and blue as the sea, has long been associated with fertility and rain. Thus it is no surprise that the maturation of the three early poets' spiritual quest coincides with the arrival of rain.

About two centuries later, the ninth-century female poet Āṇṭāḷ exploits the connection between rain, the fruition of spiritual desire, and the cultivation of virtue in her lovely thirty-verse *Tiruppāvai.* In this poem, set in the month of Mārkaḻi (December-January) that ushers in the dark part of the year, young cowherd maidens undertake a vow to win Kṛṣṇa "singing the names of the perfect one." In the third verse of the poem, the girls speak of their vow:

Singing the names
of the perfect one who
spanned the worlds with his feet
and measured them

we bathe at the break of dawn
and proclaim:

If we undertake this vow

our land will be free from evil

rains will fall three times a month
and the *kayal* will leap agilely
amidst the thick, tall, red grain

the spotted bee will sleep
nestled in the *kuvaḷai* bloom.

and when we clasp their heavy udders
the great, generous cows
will fill our pots ceaselessly

limitless wealth is certain to abound.

ēl ōr empāvāy[4]

In Āṇṭāḷ's *Tiruppāvai*, the coming of the rains, not just once but *three* times, is linked intimately to ritual practice, embodied in the singing of the names of god, here Viṣṇu and his incarnation Kṛṣṇa. Rain brings prosperity—fish leap amidst the grain, and the cows, presumably having feasted on the lush greenery brought on by a good rain, offer up more milk than the community can handle. Limitless wealth—both material and not—abounds as a result. In the *Tiruppāvai* there is no doubt that the rain *will* come, that certainty emerging from the vitality and efficacy of ritual practice. And as we have seen with the case of Poykai, Pēy, and Pūtam, when such rain comes, Viṣṇu arrives with it. In fact, in *Tiruppāvai* 4, the body of the dark monsoon cloud is compared to the body of Viṣṇu—thunder is like the resonance of his conch, the rain a shower of arrows from his fearsome bow, the bright lightning is the flash of his fiery discus. In *Tiruppāvai* 4, the innocent cowherd girls entreat the rain to fall, so that they may fulfill their quest for the divine lover. These two verses—*Tiruppāvai* 3 and 4—are intimately bound together, where the vow produces the rain, and the rain enables the vow to be fulfilled.

Yet for all this praise of the rain and certainty of its timely arrival, in contrast to the Śaiva poems, for the Vaiṣṇava poets, it is the reverse that is often desired. That is, one of the great motifs of Indic literature, which these poets are among the first to use to express mystical desire, is the coming of the rain at an undesired moment. Put another way, it is the very predictability of the monsoon rain that distresses the heroine. The situation is thus: the hero always promises to return by the monsoon, but rarely does. So even while the rain might bring general well-being to the world, to the heroine the rain signals the absence of the beloved, and ushers in personal devastation. The great Vaiṣṇava poet Nammāḻvār exploits this beautifully in his *Tiruviruttam*. In *Tiruviruttam* 68, addressing the heroine waiting for her god-lover to return:

> Her Friend Said:
>
> O girl, who is like the Vaikuṇṭha
> of the great lord
> who spanned this world surrounded by the swirling ocean
> the lovely *koṉrai* have put out their buds
> waiting for your lover's return.
>
> But they have not yet bloomed
> into dense garlands of gold
> that hang down from amidst the rich canopy of leaves.[5]

This verse falls into a poetic category called *kāla mayakkam* (the bewilderment of time). If in the Caṅkam poems, landscapes and their associated time bring with them corresponding emotional states, the *kāla mayakkam* verse is the violation of that poetic edict. The coming of the rain *ought to* signal the union of the hero and the heroine, so when the rains arrive on time, but the hero does not, the unease, the inexplicable disjunction is explained away by asserting the unpredictability of the natural world. It is the golden *ko<u>nr</u>ai* flower that has been fooled by a capricious rain into blooming at the wrong time. To suggest otherwise—that the rains have in fact come at the appointed hour, and the *ko<u>nr</u>ai* has bloomed when they ought to, asserts a certain moral failure on the part of the god, for instead of causing happiness and prosperity, he causes the opposite—the wasting away of the heroine. Indeed, Nammā<u>l</u>vār makes this very point, when he asks why the wind has turned hot in *Tiruviruttam* 5:

> Her Friend Said:
>
> At this time, in this city
> the cool breeze abandons its nature
> forgets everything, and breathes fire.
> Is it to ruin the luster of the girl
> from whose broad eyes tears drip like rain?
>
> She weeps for the cool and lovely *tulasi*
> of the one dark as the rain clouds
> whose scepter has bowed
> this one time.[6]

In a curious reversal, the very life-sustaining rain vanquishes the heroine. Her tears become the rain that destroys her body, her very self. Viṣṇu, the one dark as the rain clouds, brings no flourishing or nourishment—only a destruction, indexed by the fiery breeze. Indeed, in both these *Tiruviruttam* verses, the coming of the rains (either in the form of the heroine's tears or in a timely fashion) represents a moral failure on the part of the god, just as in the case of the Buddhist and Jain poems cited by Professor Monius. Viṣṇu's royal scepter has bent, for he has gone back on his word. Thus, in a most literal sense, the arrival of the rains in the context of the Tami<u>l</u> Vaiṣṇava literary tradition results in the very *opposite* of human flourishing.

Such reversals raise very interesting questions on the relationship between this kind of human unhappiness and disease and the ripening and flourishing of the spirit, and perhaps the intermittent fallibility of the virtue of god. While the failure of rain is read in the Śaiva poems of the Pallava period as reflecting the moral character of the king and his citizenry, it is quite the opposite in these Vaiṣṇava *bhakti* love poems. In fact, if the unimpeachable chastity of a woman is essential to the timely rains, then how do we understand the violation of a woman's chastity in the *bhakti* poems by their very arrival? No poet says it more directly than Āṇṭāḷ in *Nācciyār Tirumoḻi* 8.1:

> O clouds spread like blue cloth
> across the vast sky
> Has Tirumāl my beautiful lord
>   of Vēṅkaṭam, where cool streams leap
> come with you?
> My tears gather and spill between my breasts
> like waterfalls.
> He has destroyed my womanhood
> How does this bring him pride?[7]

Concluding Thoughts

One cannot write of the Pallavas and of *bhakti* poets without acknowledging the presence of temples, which kings commissioned and of which the poets sang. A Pallava site by the bustling city of Trichy is to be found in the rock outcropping known today as Rockfort. Located right by the great Kāviri River, it is home to a bustling temple to Gaṇeśa located at its very peak. Once one manages to clamber all the way to the top, what lies below is the rich verdant paddy plain of Tamil Nadu, and far off into the distance rise two fabulous temple towers. Nestled in an island embraced by two loving arms of the Kāviri, is Sriragam, where Viṣṇu reclines on his serpent, dreaming the worlds into existence. And right by Viṣṇu's great temple, is Tiruvanaikkaval, where Śiva manifests in the form of a *liṅga* of water. Situated under a Jambu tree, during the rainy season the *liṅga* is bathed in the waters of the stream in which it sits. Here are two ways of cultivating the natural, terrestrial world into a divine one, where the natural world is subjugated for a divine purpose, while simultaneously expressing the inherent divinity of the terrestrial realm. Srirangam is the *bhū loka* Vaikuṇṭha (a heaven on earth), and Tiruvanaikkaval with its miraculous *liṅga* of water is no less divine. It is perhaps a logical exploitation of the natural landscape, which the Pallavas so adroitly managed, and most spectacularly realized in the polyvalent "Great Penance Panel" rock face that Monius has discussed at length in her essay.

I want to end by returning to the idea of the cultivation of virtue, morality, and chastity as it relates to the land, and the ways in which we might understand it in the contemporary period. One such way of doing so, as seen with the Pallavas, is temple building. What of the *real* cultivation of the land? Does it retain the same kind of ambiguity one might discern in the poetical use of landscape in the Caṅkam-period poems and in the *bhakti* poems? In Anand Pandian's wonderful book, *Crooked Stalks: The Cultivation of Virtue in South India*, he takes up the case of the *kaḷḷars*, a notoriously rowdy caste group in Tamil Nadu, who were classified by the British as "robbers and thieves."[8] Under missionary influence and then under the watchful eye of the state, these *kaḷḷars* have turned to the cultivation of the soil, seeing in that process the cultivation of virtue, as exemplified in the restraint and control of their natural selves, which they see as tending toward robbery and rowdiness. Obviously the idea of the cultivation of virtue and the cultivation of the soil is not new, for Appar, the seventh-century Śaiva poet, said:

> Plow the field with true faith.
> Sow the seed of love, water it with patience,
> and pull out falsehood's weeds.[9]
>
> *Appar Tēvāram* 4.76

Such emphasis on agrarian civility, Pandian argues, suggests that Tamiḻs find something intrinsically virtuous in the practice of agriculture. As the rains fail (given the questionable virtue of Indian politicians, perhaps there is something to be said about the links between a king and the rain he brings), and the pressures of globalization mount, the *kaḷḷars* and others like them, increasingly abandon their fields and that notion of agrarian civility.

Glossary of Selected Words

Term (English)	***Explanation***
bhakti	means "devotion" and refers to an intense personal and emotional relationship with the Divine
Caṅkam (Sangam)	classic Tamil̲ literature from before the fourth century
Cen̲n̲ai (Chennai)	city of more than six million people in northeast Tamil Nadu, formerly known as Madras
Gaṇeśa (Ganesh or Ganesha)	elephant-headed god, son of Śiva and Pārvatī
Kāñcīpuram (Kanchipuram)	a holy city in South Indian Hinduism, in the northeast corner of the Indian state of Tamil Nadu (about seventy kilometers southwest of Cen̲n̲ai), capital of Pallava dynasty
Kṛṣṇa (Krishna)	a human-divine incarnation of Viṣṇu
liṅga (lingam)	a phallic shape associated with Śiva as his aniconic symbol
Pārvatī	Śiva's consort, also known as Umā
Śaiva (Shaiva)	devotee of Śiva, relating to Śiva
Śiva	one of the primary supreme divinities in the Hindu pantheon
Tamil̲ (Tamil)	Dravidian language spoken in South India, Sri Lanka, and Southeast Asia; also the related culture and people
Tamil Nadu	contemporary state in southeastern India, capital Cen̲n̲ai
Viṣṇu (Vishnu)	one of the primary supreme divinities in the Hindu pantheon

Notes

1. Editors' note: We offer this brief note to orient general readers who may not be familiar with the traditions and history of South Asia. At the end of the essay, there is a brief glossary of words, some of which may be familiar to the reader in their English renditions and not in the Sanskrit, such as Vishnu rather than Viṣṇu. As with Professor Monius's paper, this paper discusses the cultural and literary traditions of northern Tamil Nadu (see the first note of that paper for more on Tamil Nadu). For an overview of Hinduism, the religion most discussed in this paper, see Gavin Flood, *An Introduction to Hinduism* (Cambridge: Cambridge University Press, 1996). For more on Hinduism and ecology, see the book by that name, edited by Christopher Key Chapple and Mary Evelyn Tucker (Cambridge, MA: Center for the Study of World Religions, 2000). For more on the poetry discussed in this essay, see Norman Cutler, *Songs of Experience: The Poetics of Tamil Devotion* (Bloomington: Indiana University Press, 1987), and A. K. Ramanujan's seminal *Hymns for the Drowning: Poems for Visnu by Nammalvar* (Princeton: Princeton University Press, 1981).
2. Tēvakkulattār, *Kuṟuntokai* 3, *Kuṟuntokai.* Unpublished translation by the author.
3. A. K. Ramanujan, *Aiṅkuṟunūṟu* 15, *Poems of Love and War: From the Eight Anthologies and the Ten Long Poems of Classical Tamil* (New York: Columbia University Press, 1985), 95.
4. Antal, *Tiruppāvai* 3, *The Secret Garland: Translations of Antal's Tiruppavai and Nacciyar Tirumoli,* ed. and trans. Archana Venkatesan (New York: Oxford University Press, 2010), 53, by permission of Oxford University Press, Inc., www.oup.com. The refrain that concludes each verse of the *Tiruppāvai* is *ēl ōr empāvāy*. It points both to the genre of the poem (the *pāvai* song) as well as the polyvalence of the word *pāvai,* which means girl, doll, and vow. The refrain can be roughly translated as "O, our vow/girl."
5. Archana Venkatesan, *Tiruviruttam* 68, *A Hundred Measures of Time: Nammalvar's Tiruviruttam* (New Delhi: Penguin India, forthcoming).
6. Venkatesan, *Hundred Measures of Time.*
7. Venkatesan, *Secret Garland*, 168, by permission of Oxford University Press, Inc., www.oup.com.
8. Anand Pandian, *Crooked Stalks: Cultivating Virtue in South India* (Durham and London: Duke University Press, 2009).
9. Indira Viswanathan Peterson, *Appar Tēvāram* 4.76, *Poems to Siva* (Delhi: Motilal Banarasidass, 1991), 211.

Religious Values and Global Health

Arthur Kleinman and Bridget Hanna

I must begin with a disclaimer.* "Flourishing" has not been my topic. For forty years I have studied and written about suffering. Hence I am acutely aware of being out of my depth and even trespassing on a different domain of scholarship and practice. And yet, I am convinced that one of the sources of human flourishing is caregiving, which receives its emotional, moral, and religious impetus from the response to human suffering.

A little over a decade ago I had the good fortune of delivering the William James Lecture at Harvard Divinity School.[1] That was a greatly rewarding experience because it legitimated my effort to go beyond medicine and the mental health field to essay the broader topic of suffering as moral and religious experience. I built on that talk in the Tanner Lectures I delivered in 1998 at Stanford and in my 2006 book, *What Really Matters,* in order to set out a theory of lived experience as inherently moral.[2] The analytic argument went something like this: Experience is in the lived flow of interactions between people. It is moral at the core because there are precious things at stake for people in their local worlds and private lives. We are constantly advancing and responding to what really matters for us and for others. What is moral, however, is not necessarily ethical in the

---

* Editors' note: Although this paper is jointly authored by Professor Kleinman and Bridget Hanna, for readability and because Professor Kleinman delivered the lecture at the conference from which this paper is derived, this paper uses "I" to refer to Arthur Kleinman and when there is reference to Bridget Hanna uses her name.

sense of general principles because what is at stake for us may not be good for others or for the world. Our lives and our worlds are divided over a fundamental tension between the moral and the ethical. This tension is created and recreated at the differing levels of subjectivity, cultural meanings, and social experience. Caregiving, as I argued in the 2008 CSWR program and my essay in the resulting book, *Rethinking the Human*, is one of the inner *and* social things at stake that really does matter to our world and that bridges the moral and the ethical.[3] Yet most of the time we wear cultural blinders that restrict our awareness of the biological, psychological, and social processes shaping our sensibilities and actions—processes that on one side can constrain our humanity and foster unethical behaviors, and on the flip side, can foster caregiving, ethical aspiration, and other of the ingredients of personal and societal flourishing. In *What Really Matters*, I gave priority to the negative and troubling side of our divided self; in this paper, I examine the prosocial side that is experienced in certain forms of human flourishing.

We need to see flourishing, like suffering, as a human experience, not as a general principle or abstract universal. When we experience a health and/or social catastrophe in a family member, close friend, or in our own lives, we become aware of just how dangerous the world is, and we also come to appreciate the major role played by uncertainty. Each setting of danger and uncertainty spurs us to define what matters most to us. The work of defining and acting on what really matters, in the context of danger and uncertainty, humanizes a universal condition of living that can otherwise appear subject to the implacable workings of a cold and indifferent fate. And it is at this point that considering "flourishing" can help us understand the lived experience of religious and moral meanings in our lives.

Religious and moral meanings cultivate human flourishing by creating passionately felt values in our lives and our worlds that draw on deeply situated sources in our subjectivity for what we do, what we aspire to, and what we practice, in order to rework the human condition into something better—more just, more beautiful, and more useful for others. Both the coauthor of this paper, Bridget Hanna, and I are firmly convinced, based on personal experiences as well as understandings of the experiences of others, that deeply felt religious and moral values routinely undergird global health commitments. These values are the source of the impulse to give care and to protect others, especially the poor and the marginal. Yet, because these sources of global health practice are usually hidden or, better put, unvoiced, we routinely fail to understand the deepest motivations for global health, which may be illustrative of human flourishing more broadly.

This paper presents a number of notable historic and contemporary figures in global health and reflects on the religious and moral commitments that may inform their work. Before talking about the positive aspects of these religious commitments, I would like to acknowledge the long and often fraught history of the relationship between religion and public health. The use of medical missionaries as part of the colonial project has complex and often very destructive implications. Furthermore, religion in many forms has been implicated in the wars and genocides of the nineteenth and twentieth centuries, and in the political violence of the current time. However, the religious and moral impulse can lead to prosocial as well as antisocial actions. Albert Schweitzer explicitly framed his work as a medical missionary in Africa in the early part of the twentieth century as a response to the horrific things done in the name of God during the colonial period, noting:

> If all this oppression and all this sin and shame are perpetrated under the eye of the German God, or the American God, or the British God. . . . The name of Jesus has become a curse, and our Christianity—yours and mine—has become a falsehood and a disgrace, if the crimes are not atoned for in the very place where they were instigated. For every person who committed an atrocity in Jesus' name, someone must step in to help in Jesus' name; for every person who robbed, someone must bring a replacement; for everyone who cursed, someone must bless.

He further preached that

> When you speak about missions, let this be your message: We must make atonement for all the terrible crimes we read of in the newspapers. We must make atonement for the still worse ones, which we do not read about in the papers, crimes that are shrouded in the silence of the jungle night.[4]

Clearly, as Schweitzer recognized even then, religion's contribution to health projects has not been neutral. For this reason, many of the individuals mentioned here have tended to cloak the religious roots of their commitment to global health in secular language. This paper draws out some of those connections in order to explore the ways in which religion can be as productive a positive moral force as it can be destructive as an ideological one. In this paper, in speaking of religion, I am speaking not of religious institutions or religious ideologies; rather I am talking about the demand for compassion, humility, and doing good in the world that most religious traditions make of the individual. At a fundamental level, religious commitments motivate individuals to express what really matters to them in ethical terms—especially when it runs counter to the local moral experiences around them. This underlying contribution of religion is not always conscious for the individual, but it can contribute nonetheless by strengthening the ethical aspiration of the individual within his or her own moral life. It needs, therefore, to be seen as an important component of moral practice. I discuss global health here simply because it is the domain I know best.

When I was a student at Stanford in the 1960s, a wave of moral movements captivated my generation and propelled some of its members forward as exemplars of moral flourishing. I am thinking here of the antiwar movement, the civil rights movement, and the feminist movement. Except for the civil rights movement, where the connection to the black church is so visible, we are less likely to associate these movements with deep religious roots; but the antiwar movement that unsettled our society's disastrous military intervention in Vietnam had numerous ties to religion, from the nonviolence of Buddhists to the activism of many Christian and Jewish religious leaders. While I know much less about the determinants of the powerful wave of feminism at that time, perhaps it too had deeper roots in religious formations and popular practices.

There is currently a huge wave of student interest in global health all over the country and much of the world. One of the great moral movements of our time, this surge of interest in global health resembles the late-nineteenth-century movements of moral flourishing that followed the Second Great Awakening in American Christianity.[5] Historically these are represented by the medical missionary movement, the Boy Scout movement, the Young Men's Christian Association (YMCA) movement, the Red Cross movement, and the first efforts of the United States at building what we would now call global health. Indeed, although a patently secular human rights language dominates global health discourse and practice today, the public health and colonial health movements from which global health is descended were very often motivated by religious values. During this period, the quintessential organization of nonsectarian medical aid during conflicts, the Red Cross, was founded in 1863 in Switzerland by Henri Dunant, a businessman shocked by the terrible, bloody battlefield of Solferino in Italy in the Second Italian War for Independence.

The Red Cross symbol originated from the Swiss flag, a reversal of its white cross on the red field. Although the cross on the Swiss flag is derived from the Christian religious symbol of the cross, the main reasons this symbol was chosen were its widely recognized visibility and its connotation of the neutral state of Switzerland.[6] The red cross may have had particular religious connotations for its founder, who was raised in the Calvinist tradition and who founded the Swiss branch of the YMCA in his youth.[7] This commitment to both secular and religious values is visible again if we look at the influences of the founder of the American Red Cross, Clara Barton. Barton, whose work as a nurse, abolitionist, and activist made her famous in the nineteenth century, illustrates some of the effects of religious values as separate from institutional religion. Some of her personal thoughts on the role of religion in her work are available through a collection of papers at the Andover Library of Harvard Divinity School.

Barton trained as a schoolteacher, but grew up nursing her family. She began her humanitarian work by organizing behind the front lines in the American Civil War, almost as soon as the war began. She created an agency to distribute supplies and care for wounded soldiers. She was eventually put in charge of hospitals that were caring for the wounded. She continued her nursing and advocacy work in the United States, lecturing on her experience providing care during the Civil War and becoming involved in suffragette and abolitionist circles. In 1869 she went to Europe to recover from strain on her health, and there became involved with the International Committee of the Red Cross (ICRC), participating in their humanitarian work during the Franco-Prussian war. After her return, she was sent by the ICRC as an emissary to the U.S. president in order to set up an American chapter of the Red Cross, which she founded in 1881. Many in the United States did not believe that the country would ever see domestic combat again after the Civil War. Barton responded by reorienting the organization toward disaster relief in the United States and by eventually expanding the reach of the organization toward providing aid internationally.[8]

Barton's letters collected at the Andover Library reveal a commitment to religious values, rather than to an institutional church. These values became detached from the church itself, but continued to inform her work. She wrote in 1899:

> I do not know if I can claim a home in any Church, as I have never been a member of any, but my father heard Hosea Ballou preach the dedication sermon in the old Universalist Church in the town of Oxford, Worcester Co., Mass, and that church was his home . . . and that church was my Sunday abiding place. The cool shadows of the afternoon and the long, dark grass of the old burying place alongside are its memories, and I hope that the work of every day of my life had more or less to do with its principles.[9]

In 1904, sending a small cash donation to the Oxford church, she is more explicit:

> There are few people there [in Oxford] who have memories of harder church work or better church love in the old faith than I. In the later years of my life, I have done other things, worked along other lines than subscribing money to churches, although I have in other ways contributed my share, and it seems to me, dear sister, that with a little thought, the old church could get more through me than I could possibly give in a little personal contribution of a few dollars.[10]

Barton's story is interesting here because her work, while connected to her experience in the Oxford church, was transformed into a commitment to universal, transnational service for people in crisis. Her experience illustrates the ways that religion has informed, without dominating, humanitarian assistance and global health in the

Red Cross and the many organizations that have followed it. Other examples come to mind—the first generations of medical missionaries in China from the mid- to late nineteenth century quickly determined and made explicit in their words and actions that before they could save souls they had to save bodies. They built medical schools and universities in China aimed at dealing with the great health and social scourges of the time. John D. Rockefeller funded the faculty of Peking Union Medical College, which is still China's great institution of medical research, out of a religious conviction that he should do something to help the sick in a society that was then referred to as "the sick man of Asia." The founding of Yale-in-China Medical School in Hunan by Edward H. Hume also reflected missionary commitments that broadened out to rural support for helping the sick poor.[11] Similarly, the recent Jubilee 2000 movement to forgive Third World debt had as one of its sources the very same religious institutions in the United Kingdom that had supported the abolitionist cause in the nineteenth century.[12]

Two of the most important figures in developing this movement in global health during the twentieth century also came, in different ways, out of the medical missionary movement. Halfdan Mahler was the director-general of the World Health Organization (WHO) between 1973 and 1988. He is credited with launching at Alma Ata in 1979 the Global Strategy for Health for All by the Year 2000, and is legendary for his commitment to primary health care for all.[13] His career at the WHO began in India at the National Tuberculosis Program, but he started in tuberculosis with the Red Cross. His father was a Baptist minister and Mahler has called "social justice" a holy phrase.[14]

Another titan of the global health movement in the twentieth century came from a background of medical missionaries. James Grant was the head of UNICEF between 1980 and 1995. During that time he launched a "Child Survival and Development Revolution," focusing on immunization, oral rehydration, and breastfeeding, to combat what he called the "global silent emergency" of children dying from preventable diseases. It is estimated that these efforts in global health have saved tens of millions of lives.[15] Where did Grant's commitment to his work for children through UNICEF come from? James Grant's grandfather had been a medical missionary, and his father,

John Grant, also did some medical missionary work. John Grant headed the Department of Hygiene and Public Health at the Peking Union Medical College in China (where James Grant was raised) and worked to develop the low-cost medical training initiatives in rural areas that became the blueprint for the "barefoot doctors" program that Mao later propagated.[16] But what has happened to these narratives of religious roots?

Many public health policy makers today seem comfortable with the idea of a "right to health care," but less comfortable with the idea of a "right to health." Liberation theology and religious values have been important influences on those who have made the commitment to a right to health and may undergird their passions for health as a human right. The work of Paul Farmer and Jim Kim provides a striking contemporary example. As physician-anthropologists dedicated to bringing health care to the poor across the globe, they have been pioneers and fierce advocates for equal access to life-saving drugs for HIV/AIDS and tuberculosis in particular. They have helped to catalyze global health as the moral movement that we see today. Tracy Kidder's 2003 biography of Farmer, *Mountains beyond Mountains,* has inspired young people across the globe to support the global health movement as an ethical imperative.[17] Farmer and Kim's organization, Partners In Health (PIH), now directed by their equally impressive cofounder Ophelia Dahl, has branched out from their original projects in Haiti and then Peru. PIH is now working both within and outside the public sector to try to bring quality primary care and infectious disease treatment to needy populations in other countries.[18] Kim's work on the Green Light Committee at the World Health Organization has helped to drive down the cost of life-saving drugs for treatable infectious diseases, catalyzing a revolution in drug distribution that is saving millions of lives worldwide.[19]

Farmer and Kim rarely discuss religious values. Both come from a Catholic background that has influenced their work, particularly with the poor. They allude to this in the inspiration that they draw from the tradition of liberation theology and the Social Gospel. In *Pathologies of Power,* for example, Farmer quotes Brazil's Leonardo Boff, who writes that "the church's option is a *preferential option for the poor, against their poverty.*"[20] Farmer notes that "a preferential option for the poor offers both a challenge and an insight." The challenge, he writes, is for medical professionals and others to make the poor their true commitment and their first priority. The insight is that the poor are much more vulnerable to the ravages of disease than the wealthy, and that "diseases themselves make a preferential option for the poor," meaning that they are sicker and more vulnerable.[21] This insistence on a clear-eyed view of the consequences of inequity has driven Farmer's work in rural health in Haiti, and has also been central to some of his criticisms of the uses of human rights language. While liberation theology has always been about the struggle for social and economic rights, human rights language is often lethally neutral on these subjects. It addresses acts of active violence and repression, yet systematically neglects the structural, but even more deadly, violences of poverty, disease, and inequality.

The World Health Organization estimates that between 30 and 70 percent of health care in Africa is provided by faith-based organizations.[22] There have, of course, been diverse but significant contributions of faith-based care to the rest of the world as well. Though there are many ways in which faith-based care is similar to the care provided by secular organizations, there are also relevant differences. Caregiving, for example, has long been a central commitment of the work that many religious organizations have done in health, and is one supported by religious texts across belief systems. Such affective commitment to engaging the suffering of the other may have positive reverberations for the quality of health care delivery within faith-based organizations.

In addition, a complex mix of religious values and religious belief systems continues to fuel the political debates that surround support for global health initiatives. A good example is the passage of the U.S. President's Emergency Plan for AIDS Relief, or PEPFAR, which was funded by George W. Bush in 2003 and committed $15 billion over five years to fight the global HIV/AIDS pandemic. While religious groups had previously blocked such legislation, the support of several prominent evangelical leaders such as Billy Graham and Rick Warren eventually catalyzed the commitment. Despite the controversial requirement to spend one third of prevention money on abstinence-only education (which was changed when PEPFAR was renewed in 2008), PEPFAR has had an enormous effect on HIV/AIDS treatment worldwide, helping to make anti-retroviral therapy widely available.[23]

In a recent special issue of *Global Public Heath,* "Values and Moral Experience in Global Health," Kearsley Stewart, Gerry Keusch, and I argue that "values" are undertheorized in global health, in part because the way that religious thinkers and others who have helped to frame human rights claims as secular universals has actively downplayed their religious commitments in favor of a more inclusive (and not specifically religious) set of claims. Yet this does not change the fact that the moral and ethical principles that structure these claims are still drawn from rich traditions of religious and spiritual thought. This is demonstrated historically and anthropologically in a diverse body of work on the intersection between local knowledge and beliefs with the ideologies of development, global health, and human rights. In the special issue of *Global Public Health,* we try to frame and articulate the values that structure global health as a discipline, and emphasize the need to understand and reflect on what these values mean on the ground.[24] This would seem to be the work of medical ethics. Yet that field too, whose first generation of practitioners were so strongly influenced by James Gustafson's famous seminar in theology, has largely camouflaged the traces of religious origins and has shied away from the questions raised here.[25]

While I have variously theorized a distinction between personal ethics and the "local moral world," global health values present a unique challenge in negotiating between the two. Global health and human rights languages need to be both firm and flexible on the ground, drawing on and reverberating with local values and particularist religious beliefs, yet simultaneously representing, everywhere, the ideals of human equality, social justice, and a universal ethical aspiration. The neutrality of human rights language, as Barton, Farmer, and many foundational thinkers of both human rights and bioethics have understood, is crucial to its strength and universality. It is unclear, however, why this demands that the rich and dynamic frameworks of religion must be excluded from the conversation.

We usually think of global health as an example of a secular tradition of science coming together with humanitarian assistance. This conventional conceptualization draws too simplistic a distinction between the religious and the secular. As Max Weber pointed out early in the twentieth century, a Protestantization of religion also affected the liberal tradition, such that a self-authorizing subjectivity emerged that supported the major developments of modernity.[26] The liberal political and legal traditions of the West also informed the missionary and humanitarian assistance projects. In this sense, at particular moments, liberalization and religion were quite compatible. If we look at the personal histories of numerous contributors to global health, we often find this kind of conjunct impulse in their background and commitments, clearly visible in an Albert Schweitzer—less so, but still active, in a Paul Farmer.

Why is it that the culture of medicine and public health leads us to dismiss the contribution that religion makes to global health? I would like to argue that the culture of medicine leads to an obfuscation of these religious lineaments of moral action because they do not appear to fit within scientific modernity. In fact, for many whose silence on this point contributes to the failure to recognize this religious lineament to flourishing in the global health field, they fit very well. The contributions that global health makes more generally to human flourishing frequently demand a passionate commitment to social justice, health equity, and agonistic generosity to strangers that religious roots help to authorize and sustain. What gives strength and fortitude to those who do the difficult work of global health practice on the ground, where they may deal with loneliness, privation, illness, and other personal trials? I argue that it is a moral practice that draws on deep wells within us. William James saw these as the psychophysiology of religious inspiration.[27] There is something in our deep and divided subjectivity that is the emotional and physiological basis of religious aspiration and commitment. Out of it comes remarkable, life-sustaining powers of caregiving that are foundational to the ethical passion required by many in global health, humanitarian intervention, and those fostering other prosocial forms of human flourishing.

And what would happen if we did talk about it? Is there a way of fostering an environment in which we could discuss the religious foundations of personal commitments to global health and humanitarianism? Perhaps by opening the door to discussing the ways that religious roots contribute to producing the remarkable individuals that make our world so much better, we can open the door to a conversation about how religious and secular guidelines toward compassion, morality, and service can work proactively to support and inspire the next generation of global health leaders.

Here I am not thinking so much of institutionalized religion as of an inchoate religiosity that builds on the idea that what is most at stake in our lives is the sacredness of our own relationship to one another, our devout sensibility to care for the afflicted and marginal, and our powerful desire and need to do good in the world. This by no means takes away from the fact that there are many people who do global health and humanitarianism out of a secular orientation and an explicitly nonreligious way of being in the world; it only points to the fact that we have been dangerously silent about an aspect of the lived experience of a number of global health protagonists to the detriment of identifying those aspects of subjectivity that need to be encouraged for human flourishing within global health, and as a consequence of it, more generally in everyday life.

My coauthor, Bridget Hanna, and I do not belong to any institutionalized religions, albeit I was raised in Judaism and my wife over forty-five years socialized me to Protestant and Confucian values. As an anthropologist I can see how mankind may have created God(s) (though not out of nothing), whose reality is then legitimated and naturalized by the work of culture, but still I have no doubt that religious processes are at work in us and have encouraged my own limited contributions to teaching and research in the global health field, as well as in caregiving. And I hold that this is true of many others as well, including the esteemed respondent to this paper, Paul Farmer, who has convinced me by his own extraordinary practices that I am not off the mark. The examples used here come primarily from the Western tradition. But let there be no mistake, I could write the same chapter with examples drawn from Chinese and other secular and religious traditions globally. What I am arguing here relates as well to Islamic, Buddhist, and Neo-Confucian contributions to humanitarian assistance. In short, I believe this paper is identifying a universal aspect of the human condition.

Being able to function or flourish has to be understood in the context of a divided world and divided self. It is not just the self that is divided; the world is divided. Dealing with divisions has a physiology; it is not only a cognitive activity. Where William James argued for a psychophysiological basis for the religious inspiration, we also know there is a psychophysiological basis for selfish desire.[28] Anthropology frequently has been too simplistic about the self. Similarly in global health, the self tends to be typified as a simple caricature of someone who does things, makes decisions, is a rational self, finds the will, and then acts. We consider the social world that way too, yet there are so many divisions there: in politics, the divisions between agencies, between bureaucrats, and between the vision or mission of an organization and what the actual local world reality is—what people do to protect their careers and lives. We need to realize that to work in global health requires more than identifying that you want to do so or exactly what you want to do. Increasingly students are recognizing that they need to negotiate between a perspective that comes out of a tremendous passion for doing something in the world played against a perspective of "What about my career? Can I afford to do this?" I lived for six and a half years in East Asia doing research, often under considerable privation. As I look back over that time, I ask myself, how did I ever do that? What makes some people at one stage of their career able to do something that at another stage they find difficult, if not impossible? What helps those such as Paul Farmer endure?

I mention religious impulses and the pyschophysiological basis for them because I think that may be the connection of religion to what has inspired people's involvement in global health. Like Schweitzer, some reflect on scripture and philosophy—he was very theologically oriented—but for others such as Mahler and for James Grant, their religiosity is buried in their formation, in how they developed as moral individuals. At the level of psychobiology, I remember that religious impulse. In the temple as a boy of eight, nine, ten, I remember physiologically—viscerally—feeling uplifted as I sang and as I chanted certain of the Hebrew texts that I myself could not understand despite eight years of Hebrew School. The physiology moved me in a very powerful fashion. That visceral sense, the embodiment of the religious experience, is crucial for the commitments that people make. When we speak of values and of things related to values, we already are at a level different from the level of embodiment. The level of embodiment is the level of lived values: where we are not even aware ourselves of our values. If someone asked us about our values, we would have trouble, even though we could come up with some conventional saying, because that visceral embodied basis of religion is at a psychophysical level, different from reasoning or thinking about religion and values.

In this paper I have tried to discuss a physiology that works in a divided way, a structure of social relationships, and a set of cultural meanings—in the middle of all of these is caregiving and what it stands for. There have been tremendous advances in implementing organized responses to disease and in building systems. I wonder whether caregiving gets lost in that achievement. Do we end up counting bodies or counting the number of vials of anti-retroviral drugs we use without thinking about the actual practices of caring for people? Religion is particularly important in reminding us of the need to care for people. Thai Buddhism, for example, has encouraged some extraordinary examples of palliative end-of-life care. Similarly the Chinese tradition of Confucianism and Neo-Confucianism has organized an understanding of what caring is: the deeper you cultivate the self in Confucianism, the more you find the universal context of caring for others, rather than the self as partisan. As we move from the impulse to act toward actually building programs, the linking point of caregiving often gets lost. To return to first principles, which Paul Farmer suggests we do, would be to begin with caregiving.

In medicine, in anthropology, and almost every other field, we have failed to understand the primacy of caregiving: the practices of care, where they come from, what they do. Part of this is a blindness related to the fact that women, the poor, and immigrants do most of the caregiving. But part of this failure of understanding is an unwillingness to stretch from biology to cultural meanings. The impulse that I am identifying is both religious and moral; this physiology is a physiology of caregiving. I am interested in how that impulse finds expression: how it solidifies around caregiving, that particular form of doing for others.

So how is what I am calling our inner religious impulse for caregiving as a source of global health and human flourishing best conceived, understood, and explained? And what does its examination add to our gathering appreciation of why global health has become a defining moral movement of our time? I have suggested the image of a divided self to think through the dynamic, transformational processes in our subjectivity that result from the clash of the psychophysiology that underwrites caregiving with that which underpins more self-centered and selfish interests. Both are in near constant interactions with cultural meanings and social relationships that together—under the influence of political and economic power—remake experience. That remaking of experience is the story of different historical eras and different societies' local moral worlds. And as I discussed in *What Really Matters*, out of remade experience comes a remade subjectivity, a new and different personhood. Clearly religion plays a role in that remaking at several levels: a role that has both prosocial and antisocial possibilities and consequences. Here I emphasize the prosocial. The experience of a divided self and a divided local world is widespread in both authoritarian and liberal democratic societies. The image of the divided self has been depicted occasionally in art—including Huang Yongyu's *Owl (Maotouying)* and Picasso's *Head of the Medical Student.*[29] The figure in each of these images has one eye opened and one eye closed, portraying the tension arising from the moral life/ethical aspiration divide. One eye open to the world of actual moral experience, including the constraints on moral life, caregiving, and other practices crucial to flourishing; the other eye closed, not just to protect self-interest, but so that the self can aspire to ethical practices which are important for personal flourishing and prosocial action. The embodied feeling of a religious and ethical impulse is crucial to how the person acts under this divided condition. Without that emotion, that passion, perhaps no prosocial and self-actualizing action, especially those that run against the grain of the status quo, would be started. Yet out of the divided self also comes critical self-reflection, a scrutiny that is crucial to understanding the possibilities, limits, and unanticipated consequences of humanitarian acts, so that both aspects of a divided self and social world contribute to caregiving and other forms of human flourishing. The benefits and dangers of a religious sensibility emerge and are potentially controlled by this central process in human experience.

Notes

1. See Arthur Kleinman, "Everything That Really Matters: Social Suffering, Subjectivity, and the Remaking of Human Experience in a Disordering World," *Harvard Theological Review* 90, no. 3 (July 1997): 315–335.
2. Arthur Kleinman, "Experience and Its Moral Modes: Culture, Human Conditions and Disorder," in G. B. Peterson, ed., *The Tanner Lectures on Human Values* (Salt Lake City: University of Utah Press, 1999), 20: 357–420, also available on the web at http://www.tannerlectures.utah.edu. See also Arthur Kleinman, *What Really Matters: Living a Moral Life amidst Uncertainty and Danger* (Oxford: Oxford University Press, 2006).
3. See Arthur Kleinman, "Caregiving: The Divided Meaning of Being Human," in J. Michelle Molina and Donald K. Swearer, eds., *Rethinking the Human* (Cambridge: Center for the Study of World Religions at Harvard Divinity School, 2010), 17–29.
4. Both quotes are from Albert Schweitzer, "The Call to Mission," sermon preached on Sunday, January 6, 1905, at St. Nicolas' Church, Strasbourg, Alsace, in Schweitzer, *Essential Writings*, ed. James Brabazon (Maryknoll, NY: Orbis Books, 2005), 79, 80.
5. The rise in democratic religious and evangelical movements in the nineteenth century in the United States is often referred to as the Second Great Awakening (with Jonathan Edwards and others being seen as leading the First in the 1740s). See Nathan O. Hatch, *The Democratization of American Christianity* (New Haven: Yale University Press, 1989), for more information about some of these movements and the idea of the Second Great Awakening, particularly "Redefining the Second Great Awakening," 220–226. See Jon Butler, *Awash in a Sea of Faith* (Cambridge: Harvard University Press, 1990), for a somewhat different view.
6. See the website of the International Red Cross (ICRC) for "The History of the Emblems," April 1, 2007, accessible at http://www.icrc.org/web/eng/siteeng0.nsf/html/emblem-history.
7. See the biography of Dunant on the website of the Nobel Peace Prize (which he was awarded in 1901): http://nobelprize.org/nobel_prizes/peace/laureates/1901/dunant-bio.html.
8. See the Clara Barton biographical note on the website of the American Red Cross Museum (http://www.redcross.org/museum/history/claraBarton.asp) for further reference.
9. Clara Barton to Hosea Starr Ballou, 19 April 1899, Clara Barton papers, 1862–1911, Andover-Harvard Theological Library, Harvard Divinity School. Hosea Starr Ballou was the son of the grand-nephew of Hosea Ballou.
10. Clara Barton to Mrs. Jennie S. M. Nintur, 6 October 1904, Clara Barton papers, as above. Barton often addressed her female correspondents as sister; Nintur was not actually her sister.
11. For more information about the relationship between the Rockefeller Foundation and Peking Union Medical College, along with Yale-in-China, see Mary Brown Bullock, *An American Transplant: The Rockefeller Foundation and Peking Union Medical College* (Berkeley: University of California Press, 1980). For more on Hume and Yale-in-China, see Lian Xi, *The Conversion of Missionaries: Liberalism*

in American Protestant Missions in China, 1907–1932 (University Park: Penn State University Press, 1997).

12. See, for example, Paula Goldman, "From Margin to Mainstream: Jubilee 2000 and the Rising Profile of Global Poverty Issues in the United States and the United Kingdom," unpublished Ph.D. dissertation, Department of Anthropology, Harvard University, 2010.
13. The text for the Declaration of Alma-Ata from the International Conference on Primary Health Care, September 1978, is available from the website of the World Health Organization (WHO): http://who.int/publications/almaata_declaration_en.pdf. The primary health care topic page on the WHO website is helpful: www.who.int/topics/primary_health_care/en. For Mahler's thoughts thirty years later, see WHO, "Primary Health Care Comes Full Circle: An Interview with Dr. Halfdan Mahler," *Bulletin of the World Health Organization* 86, no. 10 (October 2008): 737–816, also available on the WHO website at www.who.int/bulletin/volumes/86/10/08-041008/en/index.html. For more information on the First International Conference and Mahler's role, see David A. Tejada de Rivero, "Alma-Ata Revisited," *Perspectives in Health Magazine* 8, no. 2 (2003). See also WHO, *The Third Ten Years of the World Health Organization, 1968–1977* (Geneva, Switzerland: WHO, 2008); despite the years given, it has an epilogue which covers Alma-Ata, 293–310. It is available on the WHO website.
14. See Marcos Cueto, "The Origins of Primary Health Care and Selective Primary Health Care," *American Journal of Public Health* 94, no. 11 (November 2004): 1864–1874.
15. For "silent emergency," see James P. Grant, "The State of the World's Children 1981–82: Children in Dark Times" (New York: United Nations Children's Fund, 1982), 1. The estimate of children's lives saved and some other biographical information is from Grant's biography on the UNICEF website: http://www.unicef.org/about/who/index_bio_grant.html.
16. See section on Grant in David Bornstein, *How to Change the World: Social Entrepreneurs and the Power of New Ideas* (New York: Oxford University Press, 2004), 242–255.
17. Tracy Kidder, *Mountains beyond Mountains* (New York: Random House, 2003).
18. Editors' note: Partners In Health (PIH) is the organization that Paul Farmer cofounded to deliver community-based medical care, initially in Haiti and then in many other places around the world. Further information is available on its website, www.pih.org.
19. Basic information about the Green Light Committee is available on the WHO website at http://www.who.int/tb/challenges/mdr/greenlightcommittee/en/ and in R. Gupta et al., "Increasing Transparency in Partnerships for Health—Introducing the Green Light Committee," *Tropical Medicine & International Health* 7, no. 11 (November 2002): 970–976. Jim Kim and Paul Farmer are two of several authors of an article in *Lancet*, J. S. Mukherjee et al., "Programmes and Principles in Treatment of Multidrug-Resistant Tuberculosis," *Lancet* 363, no. 9407 (February 7, 2004): 474-481.

20. Leonardo Boff, as quoted in Paul Farmer, *Pathologies of Power: Health, Human Rights, and the New War on the Poor* (Berkeley: University of California Press, 2003), 139, italicized in Farmer. For more information on Boff, see his own books and also Harvey Cox, *The Silencing of Leonardo Boff: The Vatican and the Future of World Christianity* (Oak Park, IL: Meyer-Stone Books, 1988).
21. Farmer, *Pathologies of Power,* 140.
22. See the summary of a report contracted by the World Health Organization in WHO, "Faith-based Organizations Play a Major Role in HIV/AIDS Care and Treatment in Sub-Saharan Africa," Press Note, February 8, 2007, WHO Media Center, Geneva, Switzerland. The full report by the African Religious Health Assets Programme (ARHAP), "Appreciating Assets: Mapping, Understanding, Translating and Engaging Religious Health Assets in Zambia and Lesotho" (Capetown, South Africa: ARHAP, October 2006), is downloadable from ARHAP at http://www.arhap.uct.ac.za/downloads/ARHAPWHO_entire.pdf. The specific statistics are drawn from a number of reports; an explanation can be found on page 20 and is continued in note 40.
23. See, for example, R. G. Biesma et al., "The Effects of Global Health Initiatives on Country Health Systems: A Review of the Evidence from HIV/AIDS Control," *Health Policy Plan* 24, no. 4 (July 2009): 239–52, as well as Eran Bendavid and Jayanta Bhattacharya, "The President's Emergency Plan for AIDS Relief in Africa: An Evaluation of Outcomes," *Annals of Internal Medicine* 150 (May 19, 2009): 688–695.
24. See Arthur Kleinman, Kearsley Stewart, Gerry Keusch, eds. *Global Public Health, Special Issue on Values and Moral Experience in Global Health* 5, no. 2 (March 2010). See particularly the editors' introduction, 115–121.
25. See, for example, James M. Gustafson, *The Contributions of Theology to Medical Ethics* (Milwaukee, WI: Marquette University Press, 1975).
26. Max Weber, *The Protestant Ethic and the Spirit of Capitalism [Protestantische Ethik und der Geist des Kapitalismus]* (repr., London: Routledge, 2001).
27. William James, *Varieties of Religious Experience* (1902, repr., Cambridge: Harvard University Press, 1985).
28. Ibid.
29. Eugene Wang's article about Huang Yongyu's pen-and-ink piece contains a reproduction of the drawing and is available through JSTOR, Eugene Wang, "The Winking Owl," *Critical Inquiry* 26, no. 3 (Spring 2000): 435–473. Pablo Picasso, *Head of the Medical Student* (Study for Les Demoiselles d'Avignon) is owned by the Museum of Modern Art (MOMA) in New York and can be seen on MOMA's website as well as in the museum.

Personal Efficacy and Moral Engagement in Global Health

Response to Kleinman and Hanna's *Religious Values and Global Health*

Paul Farmer

"How can we understand the inner religious impulse for caregiving as a foundation for global health and human flourishing? How does an examination of this impulse illuminate the emergence of global health as a defining moral movement of our time?" When I first read these questions in Arthur Kleinman's paper, they seemed directed at my work in particular.[1] He knows I have struggled with them as a doctor, teacher, and anthropologist, and he has been asking me similar questions for more than a quarter of a century. I will return later to his work and how it has inspired and provoked me. But let me first attempt some pragmatic answers to his questions.

As Professor Kleinman points out in his paper, we must sharpen our understanding of the caregiving impulse to reach a sociologically honest and historically deep hermeneutic of global health. To call it an "impulse" already seems to put it beyond rational discussion, but let us try to do justice to the motivation, whatever it may be labelled. Before historians consider efforts like the Red Cross and figures like Henri Dunant, Clara Barton, and Albert Schweitzer, they must grapple with the elision of religious motivation in most accounts of colonial medicine and the roots of humanitarian intervention. As Kleinman notes, erasing the religious underpinnings of these movements obfuscates the personal and ethical complexities of international engagement in times of war, disaster, and epidemic disease. Although I am not a scholar of the Red Cross and the emergence of humanitarianism, I am familiar with a growing body of scholarship that finds the religious impulse as a missing piece, and I believe any honest investigation into this complex set of social processes must address Professor Kleinman's questions. I would say this in any setting, even one where the religious roots of global health activism is not a welcome topic.

Taking a closer look at these roots makes one confront first principles. Although I am not always proud to have a Catholic background, there are times when Christian theology brings me helpful clarity: opposition to the death penalty, for example, or Pope Benedict's recent encyclical, *Caritas in Veritate,* which I read on the flight here [Boston] from Rwanda.[2] I have a certain fascination with the process whereby Karl Ratzinger, someone whom I regarded as a fundamentally conservative figure in the Catholic Church, became Pope Benedict. The office he held previously, Prefect of the Sacred Congregation for the Doctrine of the Faith, has been called by some critics the heir to the Inquisition.[3] I saw him in opposition to the liberation theologians who have inspired me since my youth, and whose work I cite often, including in my doctoral dissertation in anthropology under Professor Kleinman's supervision. However, this encyclical emphasizes social, economic, and environmental justice—all central themes of liberation theology. This is the paradox of modern Catholicism: it is not easily pegged on the right or left. Many of the most conservative figures in the Catholic Church return again and again to social and economic entitlements and the rights of the poor, fully believing as they do so that they are being consistent with themselves and their tradition. As the encyclical's second sentence reads, "Love—*caritas*—is an extraordinary force which leads people to opt for courageous and generous engagement in the field of justice and peace."[4]

That perspective dwells in the sacred texts of Christianity and many other religious traditions. By returning to first principles, we often find ourselves propelled into decency and engagement with serious problems that we might prefer to avoid. A classic instance from the Christian Bible is the parable of the Good Samaritan, a story told by Jesus in answer to the question, "Who is my neighbor?"[5] The radical posing of that question in four short words—"Who is my neighbor?"—expresses the imperative of engaging with others, especially with those who are unfamiliar and different. The scripture commands us to return to first principles, which is difficult for people of privilege. For example, it takes great courage for First World Catholics to subject the reigning system of global trade—one that is fundamentally unjust and perpetuates the suffering of billions in the developing world—to scrutiny and critique because they derive daily benefits from these very arrangements. It took courage for Latin American bishops to speak out against the inequalities and injustices of their own societies. The list goes on.

I find the corporal works of mercy among the most compelling Catholic social teachings. Most of the seven corporal works of mercy are intuitive: feed the hungry, clothe the naked, and so on.[6] Two, however, became clear to me only later in life: visiting the prisoners and burying the dead. "Visit the prisoners" led me again and again into prisons around the world, in Haiti, Russia, Kazakhstan, Azerbaijan, and, most recently, in Rwanda. It has long been accepted that the conditions of penal institutions are something of a barometer for the moral health of societies.[7] In my experience, they offer a particularly sobering glimpse of the ways it is possible for people to slip through society's cracks. For example, many inmates in Russian prisons become infected with multidrug-resistant tuberculosis (MDTRB), and still receive ineffective first-line treatment with the very drugs to which their tubercle bacilli have developed resistance. In this environment, a prison sentence is more like a death sentence. Note that Russia has the second-highest incarceration rate in the world, next to the United States. Facing grim situations like this one, the religious underpinnings of my youth have helped me trap myself into doing what is difficult but needs to be done: in this case, advocating for prisoners, who deserved, no matter what put them in prison, to have a normal chance of emerging from prison alive.

"Bury the dead" is a stern injunction in medical work because there are millions of early, and often preventable, deaths. What do we do when we fail? The pragmatics of burying the dead are overwhelming. Hiring coffin makers, thinking about where to place coffins, building a morgue—we do these things to create some dignity for the deceased. One aspect of the Nazi death camps often remarked on by survivors was the way that the dead were simply thrown out into the space between barracks, like garbage. In *Pathologies of Power,* I described an experience visiting the Community Health Worker Project in Guatemala soon after the peace accords were signed.[8] The Guatemalan workers wanted Partners In Health[9] to help fund a community mental health project. When I asked them what we could do, they said, "We're going to disinter the dead and rebury them with a proper religious ceremony." Those dead had been murdered, largely by military and paramilitary forces. Giving them a proper burial was important for the living. Nothing in my training—my professional training in anthropology and medicine—had prepared me to think about caring for those already dead, or, in this case, buried. Any preparation I had came from religious teachings, which helped me find compassion with humility for suffering.

Another influence on my impulse for caregiving was the reigning skepticism about the neutral observer among anthropology circles in the 1980s. Studied neutrality is a social fiction, borrowed from legal practice, business transactions, and natural science and then applied rather uncritically to the social sciences. Studies and restudies accused classic ethnographic work of embeddedness in and even complicity with the colonial enterprise (see Talal Asad's influential work).[10] My experiences in Haiti also kept me from the temptation of neutrality. In an article I somewhat pretentiously called "The Anthropologist Within" when I was twenty-four, I argued that neutrality is simply not an option in the face of suffering for which there is obvious remedy.[11] Instead, questions arise about how to engage effectively, since choosing to act—or indeed, choosing to remain neutral—is itself a form of engagement. I discourage my students at Harvard Medical School from reading what I wrote about the quest for personal efficacy versus the quest to do social justice work for poor people. I was, I think, mistaken: the quest for excellence and personal efficacy can be transformed into something good, including something good for the poor. Professor Kleinman's example raises this point: if poor health brought Clara Barton to Europe, why did she end up in the midst of the Franco-Prussian War? Her quest for personal efficacy pulled her from medical leave to a life dedicated to serving victims of violent conflict, epidemic disease, and natural disaster. Henri Dunant's engagement with humanitarian work was even more unpredictable: he witnessed the battle of Solferino because he had come to lobby Napoleon III for colonial business advantages and was so shocked by what he saw that he went on to found the International Red Cross.[12] But for the Red Cross to do the work it aspired to do, it needed a broad movement, and Henri Dunant or Clara Barton could only play one part.[13]

The great social movements—abolitionism in England and feminism in the United States, for example—invariably start with a group of people united around certain first principles. But it is up to the individual to develop an ability for critical self-reflection, or, to use Professor Kleinman's word, to become an "anti-hero" who courageously interrogates and sometimes resists widely held norms. This requires thinking deeply about whether accepted practices and orthodoxies align with one's own first principles, and if they do not, it means trying to do something about it.

Let me say a few words about my own timidity in writing about religion. First, the Enlightenment tradition of critique bars religion from medical schools and elsewhere in the academy because it appears neither scientific nor rational. In college, my mentors in anthropology were stoutly anti-religious. For example, Weston La Barre's giant opus on the origins of religion, *The Ghost Dance,* dismisses religion on psychoanalytic grounds, to which he added some anthropology and sociology.[14] Such attitudes are and were common in the university. I myself had at least a mild hostility to Catholicism as an undergraduate fascinated by the intersection of medicine and social science. However, as a doctoral student in anthropology, specializing in Haiti, I was expected to learn everything about the cosmology of my hosts. Most Haitians today are the descendants of African slaves, since the island's original population perished from disease and conquest in the sixteenth century. Many modern Haitians practice voodoo, as it is called in popular culture, which is a hybrid derivative of Catholicism. Haiti made me start thinking more deeply about Catholicism vis-à-vis voodoo and liberation theology.

My reservations about religious roots also rise from ambivalence about Christianity's connection to state power, especially imperial power, starting in the fourth century and continuing since. Other religions that make themselves welcome in the halls of government power are similarly suspect, for when spiritual and state authority are mixed, those who diverge from the norm are in for extra punishment, and injustice can receive an apparently moral sanction. Religiously motivated persecution and war have a long, grim history. Bertolt Brecht depicts the slow-burning effects of Europe's one hundred years of confessional bloodshed in his play *Mother Courage and Her Children.* Despite a heroic spirit, Mother Courage forfeits her first principles amid the chaos of war and finally, in the play's sobering denouement, loses the children she had striven so ingeniously and adventurously to protect. Brecht presents the corruption of conscience and basic values—in this case, the ethics of family—as an unavoidable consequence of mixing state and spiritual power.[15]

Yet radical undercurrents always resist and enliven religious orthodoxies. Dogma begets internal critique. For example, liberation theology illuminates the unnecessary suffering and structural violence that grew out of Catholicism's co-infection with doctrinal orthodoxy and secular ambition: individuals afraid to profess their faith, groups disenfranchised and segregated, whole populations persecuted and displaced for failing to observe the state religion. Liberation theologians coined the term "structural violence" in the 1960s, at about the same time as it appeared in sociology and political science, in order to challenge all people of faith to plumb the depths of religion and to recognize how power influences the spiritual life.

I would like to end with a word about Professor Kleinman. His books connect local worlds and personal experience to the larger forces I have been calling structural violence. When I first met Arthur Kleinman thirty years ago, he was a professor at the University of Washington in Seattle, and I was an undergraduate at Duke. I wrote him a letter, and he answered. He did not have to answer me, but he did. We have been friends ever since, and it was his encouragement and mentorship that led me to Harvard Medical School. At the time, he was working on books like *Patients and Healers in the Context of Culture* and *The Social Origins of Distress and Disease.*[16] His books deal with China, of course, a country where I have no experience, but their approach and conclusions carry far. They revealed to me some of the multiply determined links and divisions between the world of personal experience and the world of large-scale social forces that shape our lived experiences in mysterious ways. Professor Kleinman's work opened up for me a life in medicine and anthropology, a life spent grappling with, and often fighting against, the collisions between these worlds. For anyone in this field of work, there is no better guiding question and safeguard against unintended consequences of purposive action than his challenge about the relationship between the quest for personal efficacy—and the moral and religious principles driving this quest—and the structural distribution of rights and riches that confers dignity and decent health upon some, and one day, upon all human beings around the globe.

Notes

1. Editors' note: See the last paragraph of Kleinman and Hanna's essay, directly preceding this response in this book, for the original of these questions. This response paper was originally given as remarks directly following the lecture as given by Professor Kleinman, so although the paper was jointly authored by Arthur Kleinman and Bridget Hanna, for simplicity's sake we let stand Farmer's reference to Kleinman alone as its author.
2. Pope Benedict XVI, *Caritas in Veritate*, June 29, 2009, Encyclical Letter, Rome: the Vatican (2009). Available online at http://www.vatican.va/holy_father/benedict_xvi/encyclicals/index_en.htm. All quotes are taken from the English translation on the Vatican's site: http://www.vatican.va/holy_father/benedict_xvi/encyclicals/documents/hf_ben-xvi_enc_20090629_caritas-in-veritate_en.html.
3. See, for example, Julian Coman's "Don't Allow Kerry to Take Communion, Vatican Chief Tells US Catholic Bishops" in *The Telegraph* (July 11, 2004). Available online: http://www.telegraph.co.uk/news/worldnews/northamerica/usa/1466750/Dont-allow-Kerry-to-take-Communion-Vatican-chief-tells-US-Catholic-bishops.html..
4. Pope Benedict XVI, *Caritas in Veritate.*
5. *New Jerusalem Bible,* Luke 10:25–37. In response to the question "Who is my neighbor?" Jesus tells the parable of a Jewish man who was robbed and left half-dead by the roadside. A priest and an official from his own tribe pass by before he is helped by a Samaritan, who not only comes from a different religious group, but from one traditionally antagonistic to the Jews. Jesus ends the parable by asking his questioner who is the neighbor of the robbed man.
6. Editors' note: In later discussion, Professor Farmer named six of the seven corporal works of mercy. The one he could not remember was "Visit the sick," the one he performs almost every day. For more on the corporal (and spiritual) works of mercy, see Joseph Delaney, "Corporal and Spiritual Works of Mercy," in *The Catholic Encyclopedia,* Vol. 10 (New York: Robert Appleton Company, 1911). Available online at http://www.newadvent.org/cathen/10198d.htm.
7. This heuristic was common among eighteenth- and especially nineteenth-century social reformers. For example, Jeremy Bentham was fascinated by prison design and the potential to "correct" bad behavior, culminating in his famous *Panopticon.* See Jeremy Bentham, *Panopticon* (Preface) in Miran Bozovic, ed., *The Panopticon Writings* (London: Verso, 1995), 29–95. For a contrasting perspective that plumbs the depths of social ills, see Fyodor Dostoevsky's novels, especially *The House of the Dead* (Penguin, 1985) and *Crime and Punishment* (Signet Classics, 1968).
8. Paul Farmer, *Pathologies of Power: Health, Human Rights, and the New War on the Poor* (Berkeley: University of California Press, 2005), particularly 1–5, 255.
9. Editors' note: Partners In Health (PIH) is the organization that Paul Farmer cofounded to deliver community-based medical care initially in Haiti and then in many other places around the world. Its approach is exemplified by these

words on the vision page of its website (www.pih.org): "At its root, our mission is both medical and moral. It is based on solidarity, rather than charity alone. When a person in Peru, or Siberia, or rural Haiti falls ill, PIH uses all of the means at our disposal to make them well—from pressuring drug manufacturers, to lobbying policy makers, to providing medical care and social services. Whatever it takes. Just as we would do if a member of our own family—or we ourselves—were ill."

10. See Talal Asad, ed., *Anthropology and the Colonial Culture* (1973, repr., Amherst, NY: Prometheus Books, 1985).
11. Paul Farmer, "The Anthropologist Within," *Harvard Medical Alumni Bulletin* 59, no. 1 (1985): 23–28.
12. For more on the early history and institutionalization of the Red Cross in Europe and in America, see David P. Forsythe, *The Humanitarians: The International Committee of the Red Cross* (Cambridge: Cambridge University Press, 2005), and Foster Rhea Dulles, *The American Red Cross: A History* (New York: Harper and Brothers, 1950).
13. In fact, other early humanitarians like Florence Nightingale strongly disagreed with the early approach of the Red Cross. Dunant in part defended emergency caregiving as a way to lower future government spending on disabled soldiers. Although Nightingale also cared for injured soldiers during the Crimean War (1853–1856), she opposed anything that might make war less costly and therefore increase its frequency. The two correspondingly split over the length of intervention and the degree of engagement with warring governments. See Linda Polman, *The Crisis Caravan: What's Wrong with Humanitarian Aid?* (New York: Metropolitan Books, 2010), 4–5.
14. Weston La Barre, *The Ghost Dance: Origins of Religion* (1970, repr., Long Grove, IL: Waveland Press, 1990).
15. I have written more on Brecht's *Mother Courage* elsewhere; see "Mother Courage and the Costs of War" (chapter 19) in Haun Saussy, ed., *Partner to the Poor: A Paul Farmer Reader* (Berkeley: University of California Press, 2010).
16. Arthur Kleinman, *Patients and Healers in the Context of Culture: An Exploration of the Borderland between Anthropology, Medicine, and Psychiatry* (Berkeley: University of California Press, 1980) and *Social Origins of Distress and Disease: Depression and Neurasthenia in Modern China* (New Haven: Yale University Press, 1986).

A World in Crisis: The Relevance of Spiritual and Moral Principles

Chandra Muzaffar

The Crises

The world is faced with multiple interrelated crises, of which the significance of seven in particular should be emphasized. With the exception of the first two—the environmental crisis and the economic crisis—many people may not even acknowledge the other five as crises. However, if a crisis is defined as a time of danger or great difficulty, these should be classified and responded to as such.

The Environmental Crisis

An important segment of the human family has awakened to the environmental crisis. The dire consequences of global warming are registered on the radar screens of not only environmentalists but also politicians, economic planners, and media practitioners. How global warming affects climate change, vegetation, agricultural patterns, and human habitats has been the subject of numerous studies. However, efforts at curbing carbon emissions that cause global warming at both national and global levels have been weak and feeble. Political economies are too deeply mired in an economic and industrial system dependent on the burning of fossil fuels.[1]

A related dimension of the environmental crisis is the rapid depletion of nonrenewable resources. A view gaining wide currency is that (in the words of Jared Diamond) "known and likely reserves of readily accessible oil and natural gas will last only a few more decades."[2]

What happens to our civilization when we run out of these resources? Add to this prediction the rate at which we are destroying genetic diversity and wild species and losing soil fertility and we have an idea of the catastrophe that awaits future generations, unless present trends are reversed.

The Economic Crisis

The environmental crisis cannot be separated from the economic crisis since the use of renewable and nonrenewable resources relates to an economic system clearly biased toward the upper echelons of society. The widening chasm between those who have a lot and those who have a little within nation-states and across the globe is one of the gravest challenges facing humankind today. It has been estimated that the "top 10 percent of the world's people possess 84 percent of the world's wealth, while the rest are left with the remaining 16 percent."[3] In the words of the President of the Sixty-Third Session of the United Nations General Assembly, more than three billion people

> live on less than $2.50 a day. Of these, about 44 per cent survive on less than $1.25 a day, according to a new World Bank report issued on 2 September 2008. Every day, more than 30,000 people die of malnutrition, avoidable diseases and hunger. Some 85 per cent of them are children under the age of 5.[4]

If the chasm between the have-a-lot and the have-a-little has become even wider in recent years and if abject poverty continues to cripple the lives of millions of human beings especially in the Global South, it is partly because since the mid-1980s, the emphasis has been on the liberalization of financial markets, the deregulation of economies, and the privatization of public goods. This trend in global and national economies, euphemistically described as neoliberal capitalism, clearly benefits the few at the expense of the many. Neoliberal capitalism can be traced back to the abrogation of the Breton Woods system by the United States in 1971 and the decision to leave the determination of exchange rates to markets in 1973. The resulting market volatility opened the door to massive speculation, accelerated by the computer revolution and the expanding reservoir of capital available from the 1980s onwards. A powerful speculative dimension has also come to dominate contemporary capitalism.[5] More than 90 percent of global financial transactions are linked, in one way or another, to speculative capital. Speculation has triggered the rapid exit of capital from markets, ruined many an economy, and left millions of people destitute. In recent years we have witnessed the suffering it has caused to the poor in countries such as Indonesia and Argentina.

The "sub-prime mortgage crack in the U.S. housing market during the summer of 2007" which led to "the collapse of major banking institutions, precipitous falls on stock markets across the world and a credit freeze"[6] was a consequence of rampant speculation in capital and currency markets with roots in banking deregulation and excessive liquidity creation. Tens of thousands of Americans who have lost their jobs have now become the latest victims of unfettered "casino" capitalism. Cruel capitalism of this sort, pandering to the greed of a few, convinces men and women of conscience that there has to be an alternative to the present economic system.

Politics and Power

There is an intimate nexus between politics and power and economic and environmental issues, since political elites often determine priorities and formulate policies impacting directly or indirectly upon the economy and the environment. In most cases, their priorities and policies reflect the prevailing power structures, which embody the interests of the upper echelons of society rather than those of the poorer sections of the populace, in spite of the spread of democracy in recent times.[7] While it may be marginally easier to hold political leaders accountable today than a few decades ago, the interests of the upper echelons—specifically the elites—still preponderate over the rest of society.

Because political elites and their interests continue to dominate the landscape, abuse of power and corruption remain formidable challenges in many parts of the world. Even in Britain, the "mother of democracies," recent revelations of abuse and misuse of power among cabinet ministers and legislators underscore once again the importance of ethics in public life.[8]

At the global level, the situation is infinitely much worse. In the name of promoting democracy, the U.S. elite has for a long while abused its power by invading other lands, usurping resources, and eliminating those who stand in the way of the elite's pursuit of global hegemony. This hegemonic power has often flouted international law, denigrated international institutions, and ignored global public opinion with impunity.[9] If the United States has gotten away with such arrogant behavior, it is mainly because democracy has not been institutionalized in international affairs.

Military Might and Security

Might rules over right in global politics. This dictum is mirrored in the global military power of the United States. Hundreds of U.S. military bases gird the globe.[10] Its military power extends from the depths of the ocean to the outer reaches of space. There has never been a military power like the United States in the history of the human race.

It should come as no surprise therefore that the United States accounted for 58 percent of the global increase in military expenditure between 1999 and 2008. Worldwide military expenditure in 2008 stood at an estimated 1.46 trillion U.S. dollars. This represents a 4 percent increase in real terms compared to 2007 and a 45 percent increase since 1999. China, Russia, India, Saudi Arabia, Iran, Israel, Brazil, South Korea, Algeria, and Britain have also increased their military expenditures over the last decade.[11]

Burgeoning military budgets have not brought about more security. Military conflicts continue unabated in many parts of the world. Terrorism, used as the justification for expanding military expenditure in many instances, remains a major threat to the human family. Military strikes by the United States and its allies appear to have spawned more terrorist operations in places like Iraq, Afghanistan, and Pakistan.[12]

Media and Popular Consciousness

If the truth about escalating military expenditure and its contrast with the neglect of the world's poor is not widely known, it is largely because the mainstream print and electronic media have not highlighted that contrast. As part of the power structure, the media have a crucial role in preserving and perpetuating its influence and impact.[13] The emergence of cybermedia—some of which challenge global hegemony—has not changed the overall pattern of power.

Instead of consciously developing awareness about the danger of global hegemony with all its ramifications and relationship to other global crises, the media everywhere appears to encourage the growth of a superficial culture glorifying personalities from the world of sports, entertainment, and politics that relies on and promotes conspicuous consumption. The media, which derive so much of their revenue from advertisements placed by companies promoting goods and services, disseminate a global culture of consumerism.

The Culture of Consumerism

Consumerism, it has been said, is the credo of our time. By consumerism I do not mean the fundamental human need to consume in order to survive, and indeed to flourish. Legitimate consumption goes beyond the rudimentary necessities of life and encompasses genuine wants and desires that bring joy and happiness to the human being.

By consumerism I mean unrestrained consumption of goods and services, transgressing the norms of moderate spending which considers the well-being of the general public. In concrete language, possessing two or three pairs of shoes is not a problem. But why should one own twenty or thirty pairs of shoes at one time? A one-hundred dollar wristwatch may not raise eyebrows; but a ten-thousand dollar watch is certainly an example of conspicuous consumption. The "shop until you drop" obsession with buying and buying is disturbing, along with extravagance and opulence, with lifestyles that have jettisoned restraint and moderation. The conspicuous consumption of the upper class and a segment of the middle class in both the Global North and the Global South is obscene. It is greed, plain and simple.[14] It is a manifestation of narcissism, of the ego, of self-centeredness. Furthermore, the prevailing culture of consumerism is partly responsible for the environmental crisis, linked to the unequal distribution of economic and political power, skewing the allocation of resources against the poor and powerless, and supporting the hegemonic power of the wealthy elites.

Relations between Cultures, Religions, and Civilizations

Global hegemony is one of the barriers to harmonious relations between cultures, religions, and civilizations, especially in the interface between the centers of power in the West and the Muslim world.[15] Because of attempts, notably by the U.S. elite, to control and dominate the Middle East—the world's biggest exporter of oil—many Muslims are critical of U.S. foreign policy in the region. The invasion and occupation of Iraq, helmed by the United States, exacerbated relations between the United States and Muslims. More than the invasion and subsequent prolonged occupation of Iraq, the death of innocent noncombatants, especially women and children, has intensified negative feelings toward the U.S. government.

It is not just Iraq, or Lebanon, or Somalia, or Afghanistan—which have all witnessed American military involvement—that has created so much bad blood between the United States and the Muslim world. For many Muslims, the deepest wound inflicted upon the Muslim body politic remains the Israeli occupation of Palestine and the consequent dispossession of the Palestinian people. Since the United States and the West in general provide material and moral support to Israel, there is a widespread perception in the Muslim world that the United States is biased against the Palestinians and all Muslims.

The reaction of a fringe group within the Muslim world to these injustices has further aggravated ties between the West and Muslims. The terrorist network called Al-Qaeda whose tentacles stretch from Indonesia and Pakistan to Germany and Morocco has distorted Islamic teachings to legitimize its nefarious agenda of murder and mayhem. Its activities have reinforced the image of the Muslim as a terrorist in the eyes of many ordinary Americans and Europeans.

Equating Muslims and Islam with violence and terror has a long history, part of the deep-seated stereotyping of the religion and its adherents in Western Christian circles for over a thousand years. Muslim conquest of parts of the West from the eighth century onwards, the European crusades against Muslims from the eleventh to the thirteenth centuries, and Western colonialism have all contributed to this prejudice. Negative stereotypes about the West and sweeping generalizations about Christians and Jews are also rife in the Muslim world, more pervasive today than in the past. Given this background, it is not difficult to see a profound crisis in the relations between the centers of power in the West and the Muslim world.

Fundamental Causes

Global Hegemony

Global hegemony is undoubtedly one of the fundamental causes responsible for the seven crises discussed above. In an increasingly globalized world, it would be foolish to ignore or to downplay its significance.[16]

Global hegemony has played a role in wars and death and destruction. Global hegemony is linked to overwhelming global military power with all its catastrophic consequences for humankind. Global terrorism is one such consequence since it is to a large extent a reaction to occupation, a manifestation of hegemony. Global hegemony has thwarted the emergence of global democracy and accountability and has a mutually reinforcing relationship with the mainstream global media. The media have made hegemony appear normal and natural. Because of the economic forces behind global hegemony and the vested interests related to them, disparities in wealth and income between the global rich and the global poor have widened considerably. Even the culture of consumerism has become pervasive because of global hegemony. Consumerism, in turn, impacts upon finite resources and our environment. Global hegemony acts as a barrier to intercivilizational understanding and harmony.

Bias toward the Powerful and the Wealthy

But it is not just because of global hegemony that the world is in crisis. In the governance of most states throughout history, there is a bias toward the powerful and the wealthy. This bias plays a part in environmental degradation and resource depletion. Often, the interests represented by the powerful and the wealthy prevent the enforcement of effective solutions to the environmental crisis. If social and economic disparities within nation-states have increased, the wealthy and the powerful cannot be exonerated from blame. Despite electoral democracy, their dominance persists within the political arena and is often mirrored in the media. The elite from among the powerful and the wealthy decide to expand the military budget of the state, often at the expense of the well-being of the people. The powerful and the wealthy are part of the culture of conspicuous consumption and have a direct stake in perpetuating it.

Failure to Adhere to Moral Values and Principles

Apart from global hegemony and the bias toward the powerful and the wealthy within the nation-state, the failure of humankind—or more precisely, specific groups of people—to adhere to essential moral values and principles may also explain the multiple crises we face today. In each of the crises, nonadherence to certain values appears to be one of the fundamental causes. For instance, in the case of the environment, both lack of restraint and of a sense of justice are partly responsible for the environmental threats the world faces. A lack of a sense of justice also contributes to the current economic crisis, which reveals how little compassion the rich have for the poor. The abuse of power and the prevalence of corruption indicate a lack of integrity and honesty on the part of the political elite. The wanton display of military power, like the terrorist response to it, proves that both parties have scant respect for life. The media that sideline glaring facts about military expenditure, hegemony at the global level, or revelations about corruption and incompetence within the national sphere, betray their professional responsibility to tell the truth to the public. Similarly, the culture of consumerism reflects a mentality that eschews moderation and promotes an arrogant self-aggrandizement that militates against respect for cultural, religious, and civilizational diversity.

Worldview?

This brings us to the last of the fundamental causes. Unlike the other causes, this is not as readily obvious from the analyses of the crises. It is best understood by asking some searching questions about the seven crises that we have examined.

Would the environmental crisis have occurred if the dominant perspective on nature and its resources had been different—if we saw nature as part of that divine creation to which we belong, if we appreciated that transcendent link between humankind and the environment?[17] Would we have adopted such a rapacious attitude toward our finite resources if we viewed them as a temporary gift to us in trust during our brief sojourn on earth? In other words, is the environmental crisis not a consequence—to some extent at least—of a worldview from the European Enlightenment of the eighteenth century that saw nature as an object distinct from the conquering human being?

Turning to the economic crisis, would humankind have allowed such huge disparities to develop between the rich and the poor if we viewed the human family as rooted in divine unity? To achieve that unity within the human family, is it not true that justice and equality would have been the vital prerequisites? Would the global economy have become a casino driven by speculative capital if it were guided by a vision that eschewed gambling and the unrestrained accumulation of wealth and profits?

Would there be so much abuse of power in both national and global politics if politics were guided by a vision of service, if politicians regarded their power as a trust to be used judiciously in accordance with the loftiest standards of morality? Would a political leader flaunt power if he or she were deeply conscious of how puny and transient that power was in relation to the Omnipotent Power of the Almighty?[18]

Would the president or prime minister of a country conquer other lands and dominate other people, leading to the deaths of innocent children and women, if he or she genuinely believed in the sacredness of all life? Should we not expect such leaders to subscribe to that spiritual gem in Judaism and Islam that states that if you kill a single human being without just cause, it is as if you have killed the whole of humankind, and that if you save a single human being, it is as if you have saved the whole of humankind?

If everyone, especially the wealthy and powerful, acknowledged the importance of limits to consumption, if restraint and moderation were essential attributes, would we be overwhelmed by the culture of consumerism today? If our worldview were not centered upon continuously enhancing materialistic comforts, would there be greater sharing and giving to reflect a culture with a larger spiritual purpose and meaning to life?

Would a different worldview that celebrates religious and cultural diversity and respects the other have succeeded in minimizing tensions between different cultures, religions, and civilizations? Is it not because today's dominant worldview looks with fear and suspicion on differences rooted in religion, culture, ethnicity, political systems, and social customs that we have not been able to forge unity out of diversity?

Can Religion Contribute to a Solution?

Challenging the dominant worldview as a cause of today's global crises leads directly into the realm of religion. The alternative to the negative worldview discussed above is enshrined in most religions.[19] A worldview that appreciates the profound interconnection between human beings and their environment, the sacredness of life, the primacy of truth, unity within the human family, respect for the other as part of the divine gift of diversity, justice and equality as divine values, the limits of power, and the nobler purpose of life beyond consumerism, might have averted the multiple crises we face today.

Moral values such as justice and compassion, integrity and moderation, respect and restraint, so vitally necessary for human well-being, are fundamental to the various religions. The vices partly responsible for the crises—greed, self-centeredness, arrogance—are also condemned in the different religions. If the vices identified by the world's religions can be curbed and the values embodied in religious worldviews can be harnessed, we may be able to overcome the crises of our time.

If the economic, environmental, and other related crises are caused partly by a systemic bias toward those who command wealth and power, then religion can also act here as an antidote of sorts. Most religions, at the level of worldview and philosophy, are not inclined toward the powerful and the wealthy. They take to task those who abuse power and oppress others. In Christianity and Judaism, as in Islam and Buddhism, there is denunciation of those obsessed with the untrammeled accumulation of wealth. Many religions express support, sympathy, and solidarity with the poor and powerless. The lives of the founders, prophets, and great sages of the established religions bear testimony to their love and compassion for the marginalized and the poor.

Religions are opposed to hegemonic power that harms the well-being of others. The underlying principle is that no human being or institution should control or dominate another for selfish interest and gain. In Islam for instance it is through conscious, engaged surrender to God, and God alone, that human beings liberate themselves from the domination of others and the self-centeredness that drives us to try to dominate others. Muslims have consistently stood up against colonialism because the idea of liberation is rooted in the Qur'an. The first transnational movement against colonialism was initiated by an outstanding Muslim reformer, Syed Jamaluddin Assadabadi, also known as Syed Jamaluddin al-Afghani, in the latter part of the nineteenth century.[20]

Islam is not the only religion that rejects global hegemony. The values and principles contained in the basic texts of most religions show that they are averse to control and dominance by some external power. In this regard, an insightful passage from the illustrious

Chinese philosopher, Lao Tzu (fourth century BCE) establishes the philosophical basis for opposition to hegemony:

> Those who would take over the earth
> And shape it to their will
> Never, I notice, succeed.
> The earth is like a vessel so sacred
> That at the mere approach of the profane
> It is marred.
> And when they reach out their fingers it is gone.[21]

A gamut of ideas, values, and principles in the world's religions address the fundamental causes of the various crises that confront humankind. Besides, for the vast majority of the human family, religion has always been a crucial point of reference in their attempt to make sense of the vicissitudes of life. Indeed, even in societies where religion seemed to have receded from the public square, religion is resurging.[22]

Religion in Reality

Values and principles in the world's religions address the global crises I have identified, but do the millions of followers of the world's religions, and the religious leaders to whom they listen, apply these principles in constructive ways to address these crises? My response will be conditioned to some extent by the fact that I am a Muslim and it is the Muslim world with which I am most familiar. Nonetheless, I have moved beyond narrow religious boundaries: not only in the public square, but also in my private space where interacting with individuals of different religious persuasions has been an important part of my life.

To return to the question, many practicing Muslims, Buddhists, Hindus, Christians, and people of other religions see their faith as the panacea to the ills that have befallen the world. Among Muslims in particular, the rallying cry has been "Return to the Qur'an and Sunnah," in many instances augmented by the call, "Re-establish the Sharia [Islamic law]."[23] This, they feel, is the only way of overcoming the crises that confront humankind. For the advocates of a global Islamic Caliphate (a supranational entity based upon the Sharia), the return to the Qur'an and the Sunnah (way of the Prophet Muhammad) and the implementation of the Sharia have been the cornerstone of their ideology.

Since the Islamic Caliphate is a fuzzy idea, it might be more useful to examine the approaches to these crises of those Muslims who see their religion as the solution. Beginning the examination with the crisis of relations between cultures, religions, and civilizations, how have Muslim leaders addressed the present intercivilizational crisis between the West and the Muslim world, particularly since Muslims in general regard the occupation of Muslim lands and the usurpation of their resources as the root cause? Muslim religious leaders and a range of other Muslim groups have vehemently condemned U.S.-led global hegemony in general; its sometimes hypocritical postures on democracy; the propagation of what is perceived as a Western lifestyle and Western values; the global consumer culture; and what is seen as negative bias against Islam in the global media. Global economic disparities, and, in recent times, the global financial crisis, have also been subjected to trenchant criticism. On the global environmental crisis, however, Muslim religious leaders have been less vocal.

In contrast, a number of Buddhist monks and nuns in Thailand, South Korea, and Taiwan have demonstrated their commitment to the environment through various grassroots activities.[24] Some Hindu activists in India have criticized the pervasive influence of Western culture and lifestyle among the nation's middle and upper classes. Segments of Hindu society have also reacted to alleged Christian proselytization—a dimension of the crisis between cultures, religions, and civilizations. Individuals and groups within Christian orders and among lay Christians in the Philippines, South Korea, Sri Lanka, and Indonesia are not only opposed to global military and economic hegemony, but also deeply concerned about the environmental crisis and the culture of consumerism.[25]

Nonetheless, many religious leaders and lay practitioners are less concerned about the present global crises—environment degradation, gross economic inequities, spiraling militarization—than with rites and rituals, practices and traditions, and the forms and symbols of their respective religions. In other words, preserving and perpetuating—and sometimes, propagating—their religion is where the majority of religious leaders focus most of the time. Protecting their religion's identity has become their central concern in a world in crisis. Many of their followers are also committed to such a goal, especially in the global Muslim community. Muslims feel under siege; hence the need to defend their identity. Invasion and occupation and other acts of aggression of recent years have heightened Muslim insecurity and compelled a section of the community to reassert their identity. Attacks on the integrity of the community through books and cartoons and films and speeches have also made average Muslims more conscious of their identity and increasingly negative toward the West.[26]

Greater identity consciousness among Muslims—understandable under the circumstances—has nonetheless been a bane in the community, strengthening the obsession with text and tradition, dogma and dictum, prohibition and punishment, especially among the religious elites. They see it as a way of preserving the inner character and outer image of the religion. By assuming the role of protectors of the religion's identity, these elites have enhanced their power. Since the community itself looks upon these leaders as the only legitimate interpreters of the religion—the ones who decide what is permissible and what is prohibited—their moral authority is unassailable. Indeed, the interpreters of the divine word have become divine!

When dogma or practice, prohibition or punishment, has been rendered sacrosanct by religious leaders, it becomes impossible to probe text and tradition in order to understand the significance of underlying principles. And yet, these underlying principles and values should be drawn out of the Qur'an, the Sunnah, and from Islamic laws and rules as they have evolved over the centuries, to enable Muslims to formulate solutions to the multiple global crises that confront humankind. All religions must meet this challenge: to analyze texts, discard rules and edicts contradicting the universal values and principles in the philosophy of the religion itself, and then apply those values and principles to new challenges and new situations.[27]

We can address the underlying causes of some of our global crises through the application of values and principles without being bound to a specific text or tradition. As mentioned earlier, greed is one of the primary causes of the environmental crisis and the economic crisis, as well as the growth of a culture of consumerism. The Qur'an condemns greed and narrates the story of the avaricious Qarun (chapter 28:76–82). Muslim religious elites never tire of reciting the story of Qarun but they do not follow through on the implications: to curb greed one has to develop mechanisms, institute laws, and nurture a culture revolted by greed. Contemporary religious elites have failed to play a significant role in creating structures or shaping cultures to combat greed in the Muslim world.

Similarly one of the principal causes of our crises is the dominant role of hegemonic power, which many Muslims have sought to resist mainly through violence. While the Qur'an permits the victims of aggression and oppression to resist with the aid of arms, the fundamental principle of resistance itself is more important than the use of arms. Religious leaders and a significant section of the Muslim community have given more emphasis to the use of arms based upon a literal interpretation of the Qur'anic text. However, if they had grasped the significance of the principle underlying that Qur'anic teaching of resistance, they would have built instead a culture of nonviolent resistance. Peaceful, but active and dynamic resistance to global hegemony may be more effective in securing the liberation of Muslim countries under occupation. Three semi-autocratic regimes in the Muslim world—in Tehran, in Jakarta, and now in Cairo—have been overthrown by the people through peaceful means.[28] In the struggle against British colonialism in the Indian subcontinent, the Pathan chieftain, Abdul Ghaffar Khan, chose the nonviolent approach, calling it "the weapon of the Prophet, but you are not aware of it. That weapon is patience and righteousness. No power on earth can stand against it."[29]

Too often Muslim religious leaders and many ordinary Muslims are not prepared to understand and practice their faith at the level of Islam's universal principles and would rather adhere to narrow, literal interpretations of text and to simplistic legalisms in the religion. This is also true—albeit to a lesser degree—of the followers of most other religions. Since the failure to uphold the universal spiritual and moral principles and values embodied in the world's religions is one of the underlying causes of the multiple crises confronting humankind, the present approach of many Muslims to their religion is more of a liability than an asset in overcoming the crises of our time.

Transformation: A Spiritual-Moral Worldview

Can a mindset obsessed with dogma and rituals, with forms and symbols, be changed? How can such an understanding of, and approach to, Islam and other world religions be transformed? I appreciate sincerely the role of rituals and practices, forms and symbols in religious belief and practice. Prayer as a practice, for instance, has tremendous value. It links us to the Divine, to the Eternal. It reminds us of the true meaning and the real purpose of life. It strengthens our resolve to grasp and uphold the good. However, when prayer becomes primarily mechanical rite, its inner value recedes. The outward performance of prayer is too often seen as a demonstration of a person's moral worth. The rites and rituals associated with the prayer assume greater significance than the meaning and purpose of prayer, making the forms associated with a religious practice more important than its deeper meaning, purpose, and substance.

The essence of faith in monotheistic traditions is belief in God. This faith endows the gift of life with meaning, purpose, and beauty that goes beyond the self.[30] Giving substance to this belief in God makes the human being a vicegerent, a trustee—a concept embodied not only in the Abrahamic religions but also in various schools of Hindu thought and in Sikhism. As vicegerents, human beings have a mandate to protect the environment and use finite resources with great care, ever cognizant of the needs of unborn generations. As trustees, human beings are to be profoundly conscious of the sanctity of life and desist from building and perfecting weapons of death; to strive to eradicate the vast economic and social disparities that separate groups and classes and create a just and egalitarian order with the dignity of each and every person paramount; and to ensure that ethnic prejudices and stereotypes are eliminated, while celebrating religious and cultural diversity as God's cherished handiwork. Fostering a strong bond of respect and empathy among people of different faiths, and within the entire human family, is the mission of the viceregent, since unity and the peace that ensues from it are lauded virtues in the eyes of God.

Faith in a truly universal God—not a God monopolized by this or that religion—holds the promise of inspiring people to bring diverse communities together to share the joys and sorrows of life. No religious community should claim the sole possession of the Truth, that the only way to know God is through its revelation, its Prophet, or Messenger. God as the Absolute Truth should humble us as pilgrims on the journey toward that Absolute Truth beyond human comprehension.[31]

A truly universal conception of God is problematic in many different religious traditions because of the tendency to view God through a particular religious prism, colored by its own theology and history. Unfortunately, a limited theological and historical understanding of God tends to trump a universal vision. As one Zen Buddhist put it in another context, "The finger points at the moon; but one gets so engrossed in the finger that one forgets the moon." Or, in the words of the illustrious Muslim intellectual, the late Ali Shariati:

> Now we all know that religion means path, not aim; it is a road, a means. All the misfortunes that are observable in religious societies arise from the fact that religion has changed its spirit and direction; its role has changed so that religion has become an aim in itself. If you turn the road into an aim or destination—work on it, adorn it, even worship it generation after generation for hundreds of years, love it and become infatuated with it so that every time its name is mentioned or your eye glimpses it you burst into tears; if you go to war with anyone who looks askance at it, spend all your time and money on decorating, repairing and leveling it . . . if you do all of this, generation after generation, for hundreds of years, what will you become? You will become lost![32]

Instead of living a religion as a path leading to God religion becomes seen as the end in itself. We worship outer forms of religious tradition, rather than its subject, God. The transformation to which I am committed seeks to bring us back to the worship of God, and to the comprehensive universal spiritual-moral worldview associated with it.

How is such a God-centric worldview relevant to religions such as Buddhism or Confucianism when these religions do not subscribe to the belief in God? This paper cannot explore in depth how these religions stand in relationship to the concept of God. Nonetheless, a cursory glance at the subject would reveal that the Buddha himself did not repudiate the idea of God. Twentieth-century Buddhist icons like Buddhadasa Bhikku have spoken of an impersonal God who is "omnipotent, omniscient, omnipresent, eternal and absolute, thus having all the necessary qualities of the 'Supreme Thing.'"[33] In philosophical terms, scholars have argued that "Extinction (Nirvana) or 'the Void' is but God subjectivised, as a state of realization; God is but the Void objectively realized as Principle."[34] Similarly, Confucianism includes the idea of living in accordance with the Will of Heaven. Just governance rests upon the Mandate from Heaven.[35] Many ordinary Buddhists and Confucianists appeal to a reality beyond this life, a transcendence, a Divine Power that impacts upon them, and to whom they have to be accountable.

Whether or not they use the word "God," all religions seem to be grounded in a spiritual-moral dimension: spiritual because it acknowledges the Transcendent and the Divine and links the Divine to this transient existence, and moral because it bestows primacy upon values and virtues such as justice and freedom, kindness and compassion. The two are linked because the spiritual is the ultimate source of the moral while the moral derives its strength from the spiritual.

The spiritual-moral dimension within specific religions needs to be strengthened and become the basis for dialogue and interaction among different religious communities. Even though their perspectives on God or the Divine may vary, there are parallels in the way different religions accord importance to the Transcendent Reality. The parallels that are most striking are those linked to mysticism. Whether Muslim, Christian, or Hindu, mystics have adopted a universal approach to religion, in which the experience of the Divine and the substance of faith matter more than form.[36]

At the level of moral values, the similarities among the different religions are even more remarkable. They share a common moral position on the crises that confront humanity. All of them espouse living in harmony with the natural environment, establishing a moral economy, exercising political power in an ethical manner, curbing militarism, upholding the truth in the media, checking consumerism, and enhancing harmony among people of different religions and cultures. The different religious philosophies also subscribe to the view that rights cannot be separated from responsibilities.[37]

To develop a more profound commitment to the shared values and principles of various religious philosophies—a commitment from the heart, and not just the head—requires engagement with different religious groups in the larger community. People from different religious backgrounds working together on grassroots projects will develop a better appreciation of their similarities and dissimilarities. They will realize that in spite of differences, there are many commonalities in human behavior and attitude and that a common human identity transcending our distinct religious identities should be nurtured and nourished. Affirming our common human identity does not threaten our religious or other identities.

Our common humanity points to the imperative of universal spiritual-moral values and principles as we confront today's global crises: each crisis demands that all of us, whatever our religious, cultural, and ethnic affiliations, cooperate with one another. Besides, thanks to the new information and communication technologies, the borderless world is irrefutable. In the last ten years, geographical, economic, and even cultural borders have been rendered less and less important. Are we to believe that religious boundaries are somehow impervious to this monumental change?

Signs of Hope?

Are there signs suggesting that religious consciousness will undergo a significant transformation in the coming years—a transformation to weaken the hold of blind dogma and bring to the fore the universal spiritual-moral principles that undergird the world's religions? A number of developments may herald a momentous change.

One, responses to today's global crises within certain circles offer a sign of hope. Dissident groups that have raised fundamental questions about the environmental or economic or consumer crises come from different religious and cultural backgrounds but appear united in their common concerns. Even when governments adopt common positions on some of the crises through global platforms today, the dominance of a particular civilization, namely Western civilization, appears to be less overwhelming compared to a decade ago. These platforms now encompass different religions, cultures, and civilizations: the Group of 20 (rather than only the Group of 7) is one example.

Two, more significant than the wide scope of the groups and governments responding to today's crises, are underlying spiritual-moral principles articulated by the rise of alternative movements that challenge the dominant paradigm. For instance, principles such as the compassionate care of the environment, harnessed from indigenous spiritualities, are now central to the struggles of environmental nongovernmental organizations (NGOs) in both the Global North and the Global South. Similarly, the financial crisis has persuaded many in the banking and academic communities to accord much more attention to Islamic finance and its principles. Islamic finance emphasizes a secure asset foundation and discourages debt-based transactions, which played such a huge role in triggering the financial crisis. Likewise, former Iranian President Mohammed Khatami's call for a dialogue among civilizations in 1998 as a response to Samuel Huntington's "clash of civilizations" attempted to shift the focus from power to ethics in international politics. As Khatami put it, "The ultimate goal of dialogue among civilizations is not dialogue in and of itself, but attaining empathy and compassion,"[38] thus expounding a new spiritual-moral principle in international relations.

Three, the world as a whole has become much more conscious of its multireligious, multicultural character. Information technology and globalization, in general, have contributed greatly to this awareness. Members of a particular religious community with very little exposure to the religious other suddenly realize that there are millions of human beings of other faiths who share this small planet and adhere to their own beliefs and practices. This realization not only changes the community members' perception of other religions but also their understanding of the uniqueness of their own faith.

Four, globalization is forcing religious and cultural communities to go beyond mere acknowledgment of the presence of the other. A number of hitherto largely monoreligious and monocultural societies have become increasingly multireligious and multicultural as a result of the movements of peoples across borders in search of a better life. In the initial stages, the emergence of new heterogeneous environments may give rise to expressions of bigotry and antagonism, especially on the part of the host community. But with the passage of time, both host and migrant will have to learn to adjust to one another. With adjustment comes appreciation of, and even a degree of empathy for, the religion and culture of the other.

Five, besides demographic changes, globalization—its information technology dimension in particular—is impacting the popular understanding of religion in a manner whose full significance has yet to be realized, especially in the case of Islam. Websites and online commentaries now provide alternative sources of information, ideas, and analyses on controversial Islamic issues, including sources managed by religiously inclusive individuals and groups. Millions of Muslims can access such websites at the click of a button. Some of these blogs and websites have developed sizeable audiences. Many other websites and blogs continue to plug an atavistic line, but now alternative views on religious questions are widely available to a large segment of the Muslim community. This community is not confined to groups and individuals in the diaspora in Europe and America, but also embraces progressive elements in the Muslim heartland. In Indonesia, the world's most populous Muslim country, critical thinking challenging outmoded perspectives on Islamic jurisprudence has a substantial following. On a worldwide scale, such an outlook remains a subordinate trend. Nonetheless, this humane approach to Islam has strengthened a more inclusive, pluralistic understanding of the religion.

Six, if there is any group within the Muslim world that has been the vanguard of this movement toward a more critical understanding of Islam, it would be women. With emancipation through education and employment, they have begun to challenge dogmatic thinking in matters pertaining to the religion as never before. Some of these educated Muslim women—who are proud of their Muslim identity—have also mastered the Qur'an and are offering insightful interpretations about the rights of women that have exposed the misogynistic prejudices of male religious elites, past and present. These new interpretations have, by and large, buttressed the progressive dimension of Islam.

Seven, another development may also help to reinforce the progressive, universal dimension of Islam: global politics is now less dominated by the United States and the colonial legacy of Western Europe.[39] As this dominance over Muslim countries declines, the siege mentality, manifested in an obsession with identity within a significant segment of the Muslim community, may also change, allowing those Muslims to develop a more rational and sensible relationship with their identity. Dogma and prohibitions, now mobilized to define Muslim identity, may lose their grip upon the Muslim mind, making it easier for universal spiritual-moral values and principles to spread within the community.

The decline of U.S. and Western hegemonic power in general is already spawning new alliances and ventures among non-Western nations. Witness for instance the economic cooperation between states in Latin America and the Middle East; the new links forged between the African and Latin American countries; or growing trade, investment, and political ties in the East Asian region. These trends are still in the sphere of economics and politics, but they may eventually encourage the articulation of values and principles embodied in the rich spiritual and moral philosophies of the peoples of these countries.

Do all these examples, trends, and developments provide hope? They do.

But be under no illusion. For billions of children, women, and men in this world in crisis, it has been a long, dark night. The political elites who command power, the economic elites who control wealth, and the religious elites who wield influence continue to rule.

Still, we have caught a glimpse of the early glimmers, the first gleams, of light.

We await the break of dawn.

Notes

1. This is a view propounded in Wolfgang Sachs, *Planet Dialectics* (London: Zed Books, 1999).
2. See Jared Diamond, *Collapse: How Societies Choose to Fail or Succeed* (New York: Viking Press, 2005).
3. Miguel d'Escoto Brockmann, "Opening Remarks at the High-level Event on the Millennium Development Goals" (Speech, United Nations General Assembly, UN Headquarters, New York, 25 September 2008), 5, available at http://www.un.org/millenniumgoals/2008highlevel/statements.shtml. Mr. Brockman was the president of the 63rd Session of the UN General Assembly.
4. Ibid.
5. See Walden Bello, Nicola Bullard, and Kamal Malhotra, eds., *Global Finance* (London: Zed Books, 2000).
6. "World Economic Situation and Prospects 2009 Executive Summary" (New York: United Nations, January, 2009), iii. Available at http://www.un.org/en/development/desa/policy/wesp/wesp_archive/2009wesp.pdf.
7. This trend is analyzed in the context of India, the world's largest democracy, by Rajni Kothari, *Growing Amnesia* (New Delhi: Viking, 1993). See also Chandra Muzaffar, *Rights, Religion and Reform* (London: RoutledgeCurzon, 2002), especially chapter 2, "Development and Democracy in Asia."
8. See "How the Telegraph Investigation Exposed the MPs' Expenses Scandal Day by Day," *The Telegraph*, 15 May 2009, http://www.telegraph.co.uk/news/newstopics/mps-expenses/5324582/How-the-Telegraph-investigation-exposed-the-MPs-expenses-scandal-day-by-day.html.
9. See Chandra Muzaffar, *Hegemony: Justice, Peace* (Shah Alam, Malaysia: Arah Publications, 2008), especially chapters 1, 3, and 4.
10. For a detailed analysis of U.S. militarism, see Chalmers Johnson, *The Sorrows of Empire* (New York: Verso, 2004).
11. See "Global Military Expenditure Set New Record in 2008, says SIPRI" (Press Release, Stockholm: Stockholm International Peace Research Institute, 8 June 2009), http://www.sipri.org/media/pressreleases/2009/8june_yearbook_launch/view.
12. This is discussed in Abdel Bari Atwan, *The Secret History of Al-Qa'ida* (London: Abacus, 2007).
13. See Edward Herman, *Beyond Democracy: Decoding the News in an Age of Propaganda* (Boston: South End, 1992).
14. The phenomenon of greed and how it is seen in the different religions is studied in Paul Knitter and Chandra Muzaffar, eds., *Subverting Greed* (Maryknoll, NY: Orbis Books, 2002).
15. For further discussion of this point, see Chandra Muzaffar, *Global Ethic or Global Hegemony?* (London: ASEAN Academic Press, 2005), especially chapter 9.
16. The term global hegemony is analyzed in depth in Noam Chomsky, *Hegemony or Survival: America's Quest for Global Dominance* (New York: Henry Holt and Co., 2003).
17. To appreciate this argument, see Seyyed Hossein Nasr, *The Encounter of Man and Nature: The Spiritual Crisis of Modern Man* (London: G. Allen and Unwin, 1968).

18. This powerful point is made in Imam Ali, *A Selection from Nahjul Balagha* (Houston, TX: Free Islamic Literatures Incorporated, 1979), especially in the Imam's letter to the governor he had appointed, Malik-e-Ashter. Ali was the fourth, and most erudite, of the early Caliphs of Islam.
19. See Chandra Muzaffar, "The Spiritual Worldview," in *One God: Many Paths* (Penang, Malaysia: Aliran, 1980), for my first discussion of this worldview, further adumbrated in "The Spiritual Vision of the Human Being" in Chandra Muzaffar, ed., *The Human Being: Perspectives of Different Spiritual Traditions* (Penang: Aliran, 1991). Several other of my essays refine this further, the latest being "Towards a Universal Spiritual-Moral Vision of Global Justice and Peace," in Chandra Muzaffar, ed., *Religion Seeking Justice and Peace* (Penang: Penerbit Universiti Sains Malaysia, 2010).
20. See Nikki R. Keddie, *An Islamic Response to Imperialism: Political and Religious Writings of Sayyid Jamal Al-Din Al-Afghani* (Berkeley: University of California Press, 1968).
21. Quoted in Fred Dallmayr, *Small Wonder: Global Power and Its Discontents* (Lanham, MD: Rowman and Littlefield Publishers, 2005), 28.
22. Many examples of religious resurgence are provided in Peter L. Berger, ed., *The Desecularization of the World: Resurgent Religion and World Politics* (Grand Rapids, MI: William B. Eerdmans Publishing, 1999).
23. One of the best known proponents of this was the late Pakistani theologian–activist Abulala Maududi. See, for instance, his *Islamic Law and Constitution* (Lahore, Pakistan: Islamic Publications, 1955).
24. For some discussion of this, see Christopher Queen and Sallie B. King, eds., *Engaged Buddhism; Buddhist Liberation Movements in Asia* (Albany: State University of New York Press, 1996).
25. "A People's Charter on Peace for Life" (Korea: Hwacheon, 2008) expresses their philosophy.
26. Muslim identity consciousness has escalated in the last couple of decades for several reasons, discussed in Chandra Muzaffar, *Exploring Religion in Our Time* (Penang: Penerbit Universiti Sains Malaysia, 2011). See especially the chapter "Religion and Identity in a Globalising World."
27. In this regard an interesting work that examines a new methodology of interpreting the Qur'an is Abdullah Saeed's *Interpreting the Qur'an: Towards a Contemporary Approach* (London: Routledge, 2006).
28. Nonviolent resistance in Muslim settings is looked at in Chandra Muzaffar, *Muslims, Dialogue, Terror* (Petaling Jaya, Malaysia: International Movement for a Just World, 2003), especially chapters 2 and 3.
29. Quoted in Chandra Muzaffar, "A Non-Violent Struggle: The Alternative to Suicide Bombing?" in *At the Crossroads: A Malaysian Reflects on the Israeli-Palestinian Conflict* (Petaling Jaya: Bakti Ehsanmurni, 2005). For a comprehensive study of nonviolence and power, see Jonathan Schell, *The Unconquerable World* (New York: Metropolitan Books, 2003).
30. This is elaborated in Muzaffar, "Spiritual Vision of the Human Being."

31. This is a view that has been expressed on a number of occasions by the highly respected Islamic scholar, Professor Mahmoud Ayoub, who for many years was at Temple University in Philadelphia. See also the discussion of this idea of a God that is not owned by any religion in Muzaffar, *Exploring Religion.*
32. See Ali Shari'ati, *On the Sociology of Islam* (Berkeley, CA: Mizan Press, 1979), 93.
33. See Santikaro Bhikku, "Buddhadasa Bhikkhu: Life and Society Through the Natural Eyes of Voidness," in Queen and King, *Engaged Buddhism,* 161.
34. See Frithjof Schoun, *In the Tracks of Buddhism* (London: Mandala Unwin Paperbacks, 1989), 19.
35. This is explained in Huston Smith, *The Religions of Man* (New York: Harper and Row, 1958), 190–191. See also Tu Wei-ming, *Humanity and Self-Cultivation* (Berkeley, CA: Asian Humanities, 1979).
36. No mystic expressed this universal approach to religion more lucidly than the great Sufi, Jelaluddin Rumi. See *Rumi: Daylight*, trans. Camille and Kabir Helminski (Putney, VT: Threshold Books, 1990).
37. For an analysis of the rights-responsibilities nexus, see Muzaffar, "Transforming Rights: Five Challenges for the Asia-Pacific," in *Rights, Religion and Reform.*
38. See "Empathy and Compassion," *The Iranian,* 8 September 2000.
39. See in particular chapter 2 in Muzaffar, *Hegemony: Justice, Peace.*

Practice and Power
Response to Muzaffar's *A World in Crisis*

Ronald F. Thiemann

Professor Muzaffar's analysis of the seven related international crises reminds us vividly of their profound interconnectedness within our globalized world. At the center of these crises is a deep anti-democratic impulse—the accumulation of wealth and power along with the use of global violence in the hands of a few. For all the talk in the news about the "exporting of democracy," Professor Muzaffar has shown us that what is at work is in fact democracy's greatest foe: oligarchy, the accumulation of power in the hands of a few who have no accountability to the many—the *demos*—the people.

Professor Muzaffar names this form of oligarchy as "global hegemony," which fails "to adhere to essential moral values and principles." He then asks how the world might be different if these global actors had *not* adopted a rapacious self-centered worldview derived from the Enlightenment but a worldview that emphasizes limits, restraints, moderation, and diversity. What if these global actors had adopted a worldview shaped by the fundamental moral values shared by the world's major religious traditions? These are important questions, but it is precisely at this point that my worries begin to surface.

Professor Muzaffar's diagnosis is primarily about power and its misuse, but his prognosis eschews the analysis of power and focuses on the religious as worldview and values. "If the values embodied in religious worldviews can be harnessed," he writes, "we may be able to overcome the crises of our time." But what are the means whereby values are harnessed? What are the forms of counterhegemonic power that religious traditions provide which might assist us in this task? How might religious traditions work together to shape subjects and agents who embody those values and use them to resist the oligarchic powers so vividly analyzed in this presentation?

I worry, I must admit, that one could read the prognostic portion of Professor Muzaffar's paper as a *continuation* of the legacy of the European Enlightenment rather than a protest against it. Indeed, the prognosis itself seems to represent a version of classic Enlightenment liberalism of the sort that contributed mightily to the crisis he so helpfully diagnoses.

The prognosis appears to rely upon a series of distinctions all too familiar to scholars of Western religious traditions:

- form versus substance,
- performance versus sincerity,
- external versus internal,
- husk versus kernel,
- particular versus universal,
- dogma versus values.

Professor Muzaffar urges us to choose the second term of these distinctions in order to harness the powerful values inherent in religious worldviews. But these are precisely the distinctions or dichotomies that structured the unholy alliance in the modern era among liberal Protestantism, liberal political theory, and neoliberal capitalism that helped to spawn the crises so powerfully analyzed in Professor Muzaffar's paper. The preference for a values-based, universal, inwardly spiritual religion is part of a set of power negotiations that privatized religion, granted the control of violence to the state, and defined capital markets as value-free private enterprise. It is precisely that unholy alliance that must be dismantled if the hopes so poignantly expressed in this presentation have any hope of realization. And for that we have to raise the question of the kind of counterhegemonic power that might be available to religious traditions if they are to resist these death-dealing powers and contribute to the creation of ecologies of human flourishing.

While questions of values and worldviews are crucial, it is important also to ask: "How will religious traditions cultivate the subjects and agents who have the capacity for resistance and the power to construct alternate ecologies of human flourishing?" These are the questions that Dietrich Bonhoeffer, at the end of his all-too-brief life, asked as he sought to develop a seminary of resistance for the Confessing Church during the Nazi era. Central to this task was something Bonhoeffer called "the secret discipline," a series of spiritual practices derived from monastic spirituality that would create what he called a spiritual center of resistance. For Bonhoeffer spiritual practices of this sort were inherently political—external as well as internal, outwardly directed as well as inwardly directed. Bonhoeffer reminds us—as does much of the theory in the study of religion today—that subject formation is central to religious practice.[1] So we need to ask how might religious traditions—both separately and in concert—shape subjects who will enact their agencies in countercultural ways so as to resist the hegemony of global capital markets, nonrepresentative governments, and privatized religions. An emphasis on religions as communities of practice that seek to resist global hegemony is essential if the values inherent in these traditions are to be unleashed and harnessed. But that will require close attention to the forms of practice and the modes of power that shape such religious subjects and sustain religious agents against the enormous powers so well analyzed in Professor Muzaffar's paper. Attention to practice and power in addition to worldview and values is necessary for the cultivation of new ecologies of human flourishing.

Notes

1. For more on Bonhoeffer's seminary, see Eberhard Bethge, *Dietrich Bonhoeffer: Theologian, Christian, Man for His Times: A Biography,* ed. Victoria J. Barnett, rev. ed. (Minneapolis: Fortress Press, 2000), chapters 9 and 10, 419–586. For Bonhoeffer on the "secret discipline," see Dietrich Bonhoeffer, *Letters and Papers from Prison,* ed. Eberhard Bethge (New York: Simon and Schuster, 1997), particularly 286.

Cities, Climate Change, and Christianity: Religion and Sustainable Urbanism

Sallie McFague

Introduction: The Nature of Nature

In the 2007 United Nations report on climate change, the overall assessment was "unequivocal" confidence that global warming is underway and 90 percent certainty that human activities are the cause.[1] Recent updates of the 2007 report spell out an even more daunting scenario. The worst-case projections are being realized, leading to an increasing risk for abrupt and irreversible climatic shifts. In other words, the dreaded tipping point is now a realistic possibility. The twin of climate change—the economic meltdown—also faces us. These two are the products of the same insatiable desire for more: more money, more energy. Uncontrolled greed underlies both of these planetary disasters. Thomas Friedman, writing in the *New York Times* in March 2009, put it this way:

> What if the crisis of 2008 represents something much more fundamental than a deep recession? What if it's telling us that the whole growth model we created over the last 50 years is simply unsustainable economically and ecologically and that 2008 was when we hit the wall—when Mother Nature and the market both said: "No more." What if we face up to the fact that unlike the U.S. government, Mother Nature doesn't do bailouts?[2]

There are many needed responses to such news, but here I will attempt just one—an exercise in "thinking differently" about nature and our place in it, particularly in urbanized nature. Most human beings of the twenty-first century will live in cities, and cities are where half of the world's greenhouse gases are generated. What is the relationship between cities and nature and how can we achieve sustainable cities?

To be sure, a few older American cities such as New York and San Francisco, which are built high and tight, are lauded as examples of the "green metropolis," but few modern cities of ten or more million fit this category.[3] Moreover, most scholars producing theories about cities—what they are and what makes for good ones—have not paid much attention to sustainability. While many different disciplines have joined in the conversation—human geographies, social sciences, social psychology, economics, politics as well as feminist and postcolonial studies—ecological and sustainability matters have been marginalized. We postmoderns are so focused on our social constructions, our interpretations, and the products which we build (such as cities), that we forget what lies behind and beyond all our constructions, both mental and physical: it is "nature," that encompassing and mysterious term for everything that is. Catherine Keller expresses this dilemma when she writes, "Indeed, in the effort to expose the human social constructedness of the category nature we do not yet have an adequate vocabulary for naming that reality that *is* us and is *more* than us, that *something* in which we are embedded and which remains, however we (re)construct it, irreducible to us."[4] Nature is not only "irreducible to us," but it is also that upon which we rely every moment of our lives for air, water, food, and habitat. One of the consequences of our increasing awareness of human control over nature—both in thought and in action, in interpretation and in physical construction—*is the loss of a sense of our dependence on what is "more than us."* In our acknowledgement that there is no pristine nature, either in thought or reality, we have lost the sense that we are products of, sustained by, and totally dependent on "nature"—whatever we call it and whatever it is. This forgetfulness is most evident in city dwellers, because cities are "second nature," what we have built from, transformed from, changed from "first nature," this "more than us" which is never reducible to us and our constructions.[5]

If the issues contained in the threat of climate change and its twin, unsustainable economics, are to be addressed, we must overcome this forgetfulness of "first nature." We must, as the city dwellers that most of us now are, recall that "nature" is not just the trees, parks, and flowers in our cities, but, rather, it is the foundation of cities, the material from which cities are made. Every sidewalk, condo, office building, sewer pipe, electric grid, shopping mall, concert hall, parking lot, car and bus—*everything* in a city is made from nature. We do not see this lifeblood of our cities, since much of it is hidden in its new transformations—the trillions of energy exchanges that take place for every school, hospital, and jail that is built. Energy is used not only for transportation and electricity when we drive our cars or light our houses and streets. *Everything we do that involves change of any sort takes place through an exchange of energy, and energy is nothing but "first nature."* Therefore, we postmodern citizens of cities must acknowledge our situation: we are energy hogs in our use of first nature, even if we do not mean to be or are unaware of it most of the time. *We must become aware of it.* This is the challenge that cities, prime examples of "second nature," pose.

As geographer and student of cities Edward Soja rightly points out, human interpretation and manipulation do not erase nature; rather, in urbanization nature becomes a hybrid. "The urban spatiality of Nature in essence 'denaturalizes' Nature and charges it with social meaning. . . . Raw physical Nature may be naively or even divinely given to begin with, but once urban society comes into being, a new Nature is created that blends into and absorbs what existed before. One might say that the City *re-places* Nature."[6] Soja intends to undercut the conventional wisdom that views the natural and social worlds as separate, thus promoting dualism between nature and human beings. Rather, he suggests, we should focus on the hybrid—lived space—and especially on the city. By privileging space and place, everything changes, he asserts. We no longer indulge in binary dualisms of nature versus culture, but realize that at least since the beginning of agriculture and hence of urbanization, nature has never been pristine. As he puts it, we have been "without nature" for 12,000 years, which means, for all intents and purposes, always without it. There is no untouched nature, no wilderness. Even Antarctica is "urbanized," that is, socially and historically constructed.

Highlighting space is a necessary corrective to the Western and Christian emphasis on time and history and it offers several advantages. First, it focuses our attention on the earth (rather than on heaven) and does so by forcing us to attend to humanized space—cities—where most human beings now live. It spotlights the need for habitat, for humanely built spaces, here and now, rather than eternal places in heaven "by-and-by." Second, it helps us to see that natural disasters are never only natural. It forces us to accept some responsibility for the effects of global warming as well as poverty. Third, it raises issues of power and privilege in ways that a naive focus on first nature fails to do; for instance, who lives in the big house on the hill, and who lives in the shack beside the railway tracks? Fourth, it prohibits the romantic notion that all we need to do is "get back to nature," as if nature were pure, good, and available as our guide in life.

This turn to space, especially urbanized space, suggests a revolution in Western thought as well as timely attention to the space—cities—where climate change and economic sustainability will meet their harshest test. The turn to space and place—focusing on the needs of billions of human bodies as well as trillions of other creatures—is recognition that we are not robots or cyborgs, but bodies that need space. Christianity's traditional concern with time and history and its relative indifference to space and place carries Gnostic overtones. We need to remind ourselves of the following: "To be human . . . is to be placed."[7] As our planet becomes fuller with people and their increasing desire for high-energy lifestyles, *space* takes on new meaning. Everyone wants or needs more space as well as *a* place. Millions of "displaced persons" as well as cities of twelve to fifteen million human beings are the shadow side of the recent and now defunct North American real estate boom's "Location, location, location." Place, space, is about bodies and their most basic needs: food, water, habitat, medical care, education, and leisure. Time and history can bracket out these "lowly" bodily needs in order to focus on things of the mind and spirit: interpretation, meaning, and "eternal salvation."

Nature Encompasses the City

But does the turn to space and place mean that "the City *re-places* Nature"? Soja speaks of "a new Nature" that "blends into and absorbs what existed before." I fear that such language, while appreciative of the hybrid, second nature of cities, allows for and in fact encourages us to forget first nature as the source of our being. Jane Jacobs said that "without cities, we would be poor," but I would add "without nature, we would not exist."[8] If we think of the city as absorbing or replacing nature, I fear that nature's intrinsic value as well as its finite limits will be hidden from our view. What, then, will deter our appetite for unlimited, voracious over-utilization of other life-forms and earth processes? What will control our greed, the question that our ecological and economic crises are asking?

While the city in many ways has replaced nature, both concretely as the habitat of most human beings and as our own construction, nonetheless, the bottom line is that we are totally and minute-to-minute dependent on nature and its services. Nature in the first sense—"the all-encompassing source or ground of all there is"—and more specifically our own planet earth with its particular constitution of elements suitable for living things is the sine qua non. No matter how much we transform "first nature" as epitomized by the city and by our myriad interpretations, it is not infinitely malleable. The human ability to distance ourselves from first nature, both by changing it and by objectifying it, is causing a deep forgetfulness to overtake us.

This forgetfulness is epitomized in the city dweller's relationship to food. Our ability to distance ourselves from first nature is nowhere more evident than in our ignorance and denial of our total dependence on the earth with every mouthful we eat. As Michael Pollan puts it in his book tracing "the natural history of four meals" back to their roots: "All flesh is grass." Even a Twinkie or a Big Mac "begins with a particular plant growing in a specific patch of soil . . . somewhere on earth."[9] Our flesh (and the flesh that we eat) can be traced back to the grass that feeds us. "At either end of any food chain you find a biological system—a patch of soil, a human body—and the health of one is converted—literally—to the health of the other."[10] And yet, we city dwellers have forgotten this all-important piece of information: the inexorable, undeniable link between our health and the health of the planet. The connection is from body to body. "Daily, our eating turns nature into culture, transforming *the body of the world into our bodies and minds*" (italics added).[11] We need to relearn the importance of this most basic of all transformations, "For we would [then] no longer need any reminder that however we choose to feed ourselves, we eat by the grace of nature, not industry, and what we're eating is never anything more or less than *the body of the world*" (italics added).[12]

But urban dwellers seldom see this. The city is the prime example of both our greatest accomplishment and our greatest danger. Jerusalem, the city of desire and delight, is fast emerging as Babylon, the city of excessive luxury in the midst of extreme poverty. The city, which stands as the quintessential human habitation—civilized, diverse, cosmopolitan—is at the same time becoming the greatest threat to human well-being. "Of all the recognized ecological systems it is human urbanism which seems most destructive of its host."[13] Cities suck energy from near and far to allow some city dwellers to live at the highest level of comfort and convenience ever known, while many others exist in squalor. Again, I would underscore that this does not mean that we should retreat to either the country or suburbia, for these spaces are even less energy efficient for millions of human inhabitants. The city is where most of us must live and where just, sustainable living has the best chance.

Nonetheless city dwellers must attend to the judgment of the United Nations Millennium Ecosystem Assessment, the work of over 1,300 experts worldwide, which claims that we are excessive energy users—we are literally consuming the planet. This massive study, "the first attempt by the scientific community to describe and evaluate on a global scale the full range of services people desire from nature,"[14] reaches the sobering conclusion that out of twenty-four essential services provided by nature to humanity, nearly two-thirds are in decline. These services are numerous and all-encompassing, falling into three major categories: Provisioning Services (food, fiber, genetic resources, biochemicals and natural medicines, fresh water); Regulating Services (air, climate, water, erosion, disease, pest, pollination, natural hazard); and Cultural Services (spiritual and religious values, aesthetic values, recreation and ecotourism).[15] The Assessment stresses the need for consciousness-raising: "We must learn to recognize the true value of nature both in an economic sense and in the richness it provides to our lives in ways more difficult to put numbers on."[16] Such true value ranges from the taste of a cup of clean water to the sight of snow-capped mountains. We are nature's debtors and nature's lovers. Anthropologist David Harvey sums up the situation well with these sobering words:

> A strong case can be made that the humanly-induced environmental transformations now under way are larger

> scale, riskier, and more far reaching and complex in their implications (materially, spiritually, aesthetically) than ever before in human history. The quantitative shifts that have occurred in the last half of the twentieth century . . . imply a qualitative shift in environmental impacts and potential unintended consequences that requires a comparable qualitative shift in our responses and our thinking.[17]

Thinking and Acting Differently

The task before us—"a qualitative shift in our responses and thinking"—is daunting. Yet, such appears to be the overwhelming conclusion coming from all fields that study planetary health. As botanist and conservationist Peter Raven says: "It is also a fundamentally spiritual task."[18] Most scholars agree that an attitude change is needed, a shift in values at a deep level—as well as whatever their particular discipline can contribute to solving the planetary crisis. From the time of Aristotle to the eighteenth century, economics was considered a subdivision of ethics: the good life was understood to be based on such values as the common good, justice, and limits. Having lost this context for how to live on our planet and substituting the insatiable greed of market capitalism in its stead, we are now without the means to make the qualitative shift in thinking that is required. With the death of communism and the decline of socialism, Western society is left with an image of human life that is radically individualistic, diametrically opposed to how we should think of ourselves within an ecological worldview. It is impossible to imagine us acting differently—acting as "ecological citizens"—unless we internalize ecological values.[19]

One of the distinctive activities of religion is the formation of basic assumptions regarding human nature and our place in the scheme of things. As theologians widely agree, all theology is anthropology. Religious traditions educate through stories, images, and metaphors, creating in their adherents deep and often unconscious assumptions about who human beings are and how they should act. Religions are into the business of forming the imagination and thus influencing the action of people. It is at this point that the religions can make a significant contribution to the planetary crisis. For we live *within* the assumptions, the constructions, of who we think we are. As these assumptions and constructions change, so might behavior. One small contribution toward this possibility is to change the metaphor by which we think of ourselves in the world. The conventional and widely accepted metaphor of twenty-first-century market capitalism is *the individual in the machine.* Human beings are seen as subjects in an objectified world, a "thing," which is there for our use—our needs, desires, and recreation. To see ourselves this way, however, is an anomaly in human history, for until the scientific revolution of the seventeenth century, as Carolyn Merchant and others have pointed out, the earth was assumed to be alive, even as we are.[20] From the Stoics to the worldview of First Nations' peoples, including medieval Christians, the apparent organic quality of the earth was not questioned. But during the last few hundred years, it has become increasingly useful and profitable to think of the world more like a machine than a body. We think in metaphors, especially at the deepest level of our worldviews. If the machine model dominates, then we will think of the world's parts as only externally related, able to be repaired like cars with new parts substituting for faulty ones, with few consequences for the earth as a whole. With such a model in mind, it is difficult for people to see the tragedy of clear-cutting forest practices or the implications of global warming.

The "individual in the machine" model fits easily into the sensibility of city dwellers. Since most of us do not recognize "first nature" as the source of the buildings, trucks, machines, and highways we construct, thinking of the world in terms of exchangeable parts is easy. Cities do not appear to be organic entities made from the earth; rather, they have independent parts "made by human beings" that can be torn down when needed and new ones constructed. However, the "body" is reemerging across many fields of study as a basic metaphor for interpretation and action. David Harvey mentions "the extraordinary efflorescence of interest in 'the body' as a grounding for all sorts of theoretical enquiries over the last two decades or so."[21] This interest is hardly novel: from the Socratic notion of the body ("man") as the measure of all things to the Stoic metaphor of the world as a living organism and First Nations' understanding of the earth as "mother" of us all, body language has historically been central to the interpretation of our place in the scheme of things. This is true of Christianity as well. As an incarnational religion, Christianity has focused on bodily metaphors: Jesus as the incarnate God, the

Eucharist as the body and blood of Christ, the church as the body of Christ.[22] To see bodily well-being as the measure of both human and planetary well-being is so obvious that it seems strange it should need a revival, but surfacing once again is the realization that an appropriate metaphor with which to imagine our relation to the world is not *the individual in the machine,* but *bodies living with the body of the earth.*

We have seen the results of living within the machine model for several hundred years now and the verdict is overwhelmingly negative. The body as measure, as the lens through which we view the world and ourselves, changes everything. It means that human beings as bodies, dependent on other bodies and on the body of the earth, are interrelated and interdependent in infinite, mind-boggling, wonderful, and risky ways. It means that *materialism* (in the sense of what makes for bodily well-being for all humans and for the earth) becomes the measure of the good life. Combining the socialist with the ecological vision of human and planetary flourishing, it means that the good life cannot be a small percentage of individuals hoarding basic resources for their own comfort and enjoyment. Rather, if we desire to take care of ourselves, we must also take care of the world, for we are, in this metaphor, internally related and mutually dependent on all other parts of the body. The metaphor of body—not just the human body but all creaturely bodies—is a radically egalitarian measure of the good life: it claims that all bodies deserve the *basics* (food, habitat, clean air, and water).

The Organic Model and Christianity

I would like now to investigate in more detail the contribution of a Christian understanding of the organic model for twenty-first-century urban living. The Christian incarnational understanding of the God-world relationship—that the world is from the beginning loved by God and is a reflection of the divine—means that flesh, bodies, space and place, air and water, and food and habitat are all "religious" matters. The locus of attention of incarnational Christianity is *the body,* both the world as body and the bodies that compose it. The *material* focus becomes central; incarnational theology is militantly anti-Gnostic, anti-spiritualizing, anti-dualistic. The Christian incarnational focus on bodies can be seen in the two central historical streams in Christianity:

the sacramental and the prophetic, or the Catholic and the Protestant. The sacramental dimension says that the world is a reflection of God, tells us of God, and connects the earthly, bodily joys of life (beauty, love, food, music, play) with God. The prophetic dimension insists that since the world is a body, it must be fed and cared for: all parts must receive their just supply of resources, and it must be sustained for the indefinite future. While the sacramental dimension of the model encourages us to appreciate and love others—realize their worth—the prophetic dimension focuses our attention on limits—the recognition that bodies, including the body of the world, are finite. All life-forms must have food, fresh water, clean air, and a habitat. The prophetic dimension stresses the limits of all bodies, the finitude of the planet, the need for just and sustainable use of resources.

The prophetic dimension of the model—the awareness of the finitude, limits, and needs of bodies—suggests an ethic of self-limitation for twenty-first century well-off urban dwellers, so that slum dwellers may have space and place. This dimension must take center stage. *Our very survival may well rest on living within such a construction of nature—one in which second nature is constrained.* An ethic of self-limitation could also help address the issue of climate change due to excessive urban energy use.

I have taught a course on spiritual autobiography for many years; in fact, it was the first course I taught, almost forty years ago. It is about people who live lives of extraordinary love for others, especially the weak and vulnerable—folks like Teresa of Avila, John Woolman, Dietrich Bonhoeffer, Simone Weil, Mohandas K. Gandhi, Jean Vanier, Martin Luther King, Jr., Nelson Mandela, and Dorothy Day. I always gain new insight teaching the course and last year was no exception. I have been struck by a characteristic shared by many of them, the rather shocking practice of self-emptying, of what the Christian tradition has called "kenosis." The text from Philippians sums it up well: "Let the same mind be in you that was in Christ Jesus, who, though he was in the form of God, did not regard equality with God as something to be exploited, but emptied himself, taking the form of a slave, being born in human likeness. And being found in human form, he humbled himself and became obedient to the point of death—even death on a cross" (2:5–8). What an inversion this is of triumphal, imperialistic views of Christianity!

I believe self-emptying suggests an ethic for this time of climate change and financial chaos. These two related crises are the result of excess, our insatiable appetites that are literally consuming the world. We are debtors twice over—financially and ecologically. The very habits that are causing the financial crisis are also destroying the planet. We are living way beyond our means at all levels: our personal credit cards, the practices of the financial lending institutions, and the planet's resources that support all of us.

Could the crazy notion of self-emptying, a notion found in different forms in many religious traditions, be a clue to what is wrong with our way of being in the world as well as a suggestion of how we might live differently? Whether in Buddhism's release from desire by nonattachment or Christianity's admonition that to find one's life one must lose it, religions are often countercultural in their various ethics of self-denial in order that genuine fulfillment might occur. While in some religious traditions, such self-denial moves into asceticism and life-denial, this is not usually the underlying assumption.

I am thinking of John Woolman, an eighteenth-century American Quaker, who had a successful retail business and gave it up because he felt it kept him from seeing clearly something that disturbed him: slavery. He came to see how money stood in the way of perceiving injustice: people who had a lot of property and land needed slaves to maintain them (or so these folks reasoned). He saw the same problem with his own reasoning—whenever he looked at an injustice in the world he always saw it through his own eyes, his own situation, and his own possible benefit. Once he reduced his own level of prosperity, he could see the clear links between riches and oppression and move himself out of the center. He wrote: "Every degree of luxury has some connection with evil."[23] Reduction of his lifestyle gave him insight into the difference between "needs" and "wants," something our insatiable consumer culture has made it almost impossible to recognize. As an ethic for a time of climate change, Woolman suggests the clarity of perception into others "needs" that can come about through the reduction of one's own "wants." Woolman did not find such self-emptying negative or depressing; rather, he found it fulfilling. He had a dream in which he heard the words "John Woolman is dead" and realized that with his own will dead he could say with Paul that he was crucified with Christ, that Christ might live in him. We find ourselves by losing ourselves. That deeper desire is the desire for God, for nothing less will fill the hunger in us. Augustine says that we are drawn to God as a sheep is drawn to a leafy branch or a child to a handful of nuts. To empty the self is not an act of denial, but of joy, for it creates space for God to fill one's being.

What we see here is not an ascetic call for self-denial to purify ourselves or even a moral injunction to give others space to live; rather, it is more basic. It is an invitation to imitate the way God loves the world. In the Christian tradition, "kenosis" or self-emptying is a way of understanding God's actions in creation, the incarnation, and the cross. In creation, God limits the divine self, by pulling in, so to speak, to allow space for others to exist. God, who is the one in whom we live and move and have our being, does not take all the space but gives space and life to others. This is an inversion of the usual understanding of power as control; instead, power is given to others to live as diverse and valuable creatures. In the incarnation, as Paul writes in that verse from Phillipians, God "emptied himself, taking the form of a slave," substituting humility and vulnerability for our insatiable appetites. In the cross God gives of the divine self without limit to side with the poor and the oppressed. God does not take the way of the victor, but like Jesus and the temptations, rejects absolute power and imperialism for a different way. Therefore, Christian discipleship becomes a "cruciform" life, imitating the self-giving of Christ for others.

Another example of kenotic living is the French philosopher and unbaptized Catholic, Simone Weil. She lived a radical and brief life of solidarity with her poorest and often starving fellow citizens during World War II. She practiced what she called "de-creation," a form of self-emptying in which her ego diminishes as God grows in her.[24] De-creation or the death of the will is giving up control over one's life, so that God can subvert the self's exorbitant and constantly growing desires. The point is not mortification but a discipline of emptying self so that God can be all in all. To eat when and what one wants when others are starving is a symbol of control over finitude, of exceptionalism, which Weil refused to embrace. She limited the amount she ate to no more than her starving neighbors. Food is a symbol of basic physical limits, and unless we can limit our own voracious appetites, we will not be able to attend to the hunger in others—their abject suffering, both physical and emotional. Our tendency is to love others because of our needs, not theirs, our hunger, not their hunger. We want more and more in the insatiability of the consumer culture that has resulted in climate change and more recently in financial disaster.

Simone Weil says that human beings are naturally "cannibalistic": we eat instead of looking, we devour rather than paying attention, we consume other people and the planet in our search for self-fulfillment.[25] Augustine claimed something similar in his understanding of sin: voracious, lustful desire to have it all for oneself. From the twenty-first-century ecological perspective, sin is refusing to share, refusing to live in such a way that others—other people and other life-forms—can also live. For us in our time, sin is refusing to live justly and sustainably with all others on our planet, refusing to share the banquet of life.

As with Woolman, the problem as Weil understands it is the inability to see others and pay attention to their needs rather than simply our wants. The United Nations Earth Charter, a document which lays out principles for a just, sustainable planet, agrees. Its first principle reads: "Recognize that all beings are interdependent and every form of life has value regardless of its worth to human beings." An ethic of self-emptying begins with the recognition that something besides oneself exists and requires the basics of existence.

An Ethic for Daily Life and the Urban Crisis

Paying attention to others, looking not eating, is a somber, thoughtful ethic for our time of climate change. Put simply, climate change is the result of too many human beings using too much energy and taking up too much space on the planet. "Environmentalism" is not simply about maintaining green spaces in cities or national parks; rather, it is the more basic issue of energy use on a finite planet. Thus, space and energy, the basic physical needs of all creatures—a place to live and the energy to sustain life day by day—are the issues. The crisis facing us is one of geography, one of space and place and habitation. It is not about time and history and human meaning; rather, it is physical, earthly, worldly, fleshly—the basics of existence.

This "crisis" does not have the immediacy of a war or plague or tsunami. It has to do with how we live on a daily basis—the food we eat, the transportation we use, the size of the house we live in, the consumer goods we buy, the luxuries we allow ourselves, the amount of long-distance air travel we permit ourselves, and so forth. The enemy is the very ordinary life we ourselves are leading as well-off North Americans. For all its presumed innocence, this way of life, multiplied by billions of people, is both unjust to those who cannot attain this lifestyle and destructive of the very planet that supports us all.

A very different way of life is suggested by another extraordinary Christian, Dorothy Day, who identified with the abject poverty of people in the ghettoes of New York City during the Great Depression. She lived a life of joyful sharing, a form of the abundant life totally contrary to our consumer understanding. If Woolman and Weil belong to the prophetic strain in Christianity, the strain that underscores the way to God through self-emptying, Day belongs to the sacramental path that while acknowledging self-emptying, revels in the fulfillment that follows. She found the abundant life in voluntary poverty. She did indeed find her life by losing it, and it was a rich, full, joyful life. In the postscript to her autobiography, she writes of her community:

> We were just sitting there talking when lines of people began to form, saying, "We need bread." We could not say, "Go, be thou filled." If there were six small loaves and a few fishes, we had to divide them. There was always bread. . . . There is always room for one more; each of us will have a little less. . . . We cannot love God unless we love each other, and to love we must know each other. We know Him in the breaking of bread, and we know each other in the breaking of bread, and we are not alone any more. Heaven is a banquet and life is a banquet, too, even with a crust, where there is companionship.[26]

The kenotic paradigm in Woolman, Weil, and Day includes the recognition that life's flourishing on earth demands certain limitations and sacrifices at physical and emotional levels. Our time is characterized by a growing awareness of our radical interdependence on all other life-forms and an increasing appreciation of the planet's finitude and vulnerability. These realities mean that the vocabulary and sensibility of self-limitation, egolessness, sharing, giving space to others, and limiting our energy use, no longer sound like a special language for the saints, but rather, like an ethic for all of us. The religions may be the greatest "realists," with their intuitive appreciation for self-emptying and self-limitation as a way not only to personal fulfillment but also to sane planetary practice. Could religions take the lead in exploring and illustrating how an ethic of self-limitation might function in light of the twenty-first-century crisis of climate change? The banquet of which Dorothy Day speaks—the banquet of heaven and the banquet of earth—is an inclusive feast with "room for one more" if each of us has "a little less."

So what does this have to do with cities? Recent UN projections on the growth of cities claim that slum living is now among the fastest-growing legacies of "civilization."[27] Given present trends, one out of three city dwellers will be doomed to the slums. The conditions in many cities—those pushing fifteen million—are already dire. The needs of doubling city populations by mid-century from the present two billion to four billion are mind-boggling.[28] Housing, public health, transportation, energy, food and water, education, and medical services are simply the basics for minimal human existence.

For well-off city dwellers, the kenotic, prophetic sensibility means some concrete, empirical on-the-ground changes. It means that second nature, the built environment, must be minimalized rather than maximalized. It means small condos and apartments, not mansions; living spaces that go up, not out; small, hybrid cars, not Hummers; food that is grown locally, not halfway around the world. It means saying NO, saying "enough." Second nature is built upon first nature and first nature is, increasingly, a vulnerable, deteriorating body unable to support the Western high-energy lifestyle. This realization should impact us at all levels: what we eat, our means of transportation, what we wear, the places we live, the parks where we play, the offices where we work. One of the greatest challenges of the twenty-first century is decent, livable conditions for the billions who will live in cities. We well-off city dwellers need to take up less space, use less energy, lower our desires for more, attend to "needs" before "wants"—become small, in other words. The prophetic, kenotic sensibility demands that prosperous urban dwellers retreat from expansion and accept simplification at all levels of existence. Justice and sustainability demand that whatever we build upon first nature be shared with all other beings and be done within the limits of the planet's resources.

We need to imagine living within a "bounded economy," living with restraint. The prophetic, kenotic sensibility is not just for individuals; rather, it is a characteristic of our planet. In a recent issue of the journal *Nature*, scientists named nine key "planetary boundaries" that must be respected to avoid catastrophic environmental disaster.[29] Three of these boundaries—climate change, biological diversity, and nitrogen and phosphorus inputs to the biosphere and oceans—have already been breached through human activity. The planet has limits; it demands we live within these limits.

Conclusion

Second nature needs to acknowledge the base on which it is built and will continue to depend—first nature. Totalizing theories that eliminate first nature forget that we are *bodies* before all else. We cannot interpret or build unless we can eat. While we are also interpreters and builders of our world, we are not its maker or master. There is something outside of all our interpretations and constructions: the air we must breathe, the water we must drink, and the food we must eat. An understanding of second nature that underestimates the inescapable importance of first nature is not only unhelpful in our planetary crisis—it is also false. It does not help us to live as best we can, making decisions that are relatively better, even if Eden—first nature—is no longer available. An organic model of the world reminds us of the sacredness, beauty, and importance of first nature at both the local and planetary level. It reminds us that our planet is a limited physical entity able to support millions of species as well as human beings, but only on a just and sustainable basis.

The God who opened up space for creation and who became empty in the incarnation is far removed from the image of an absolute, unmoved ruler who controls others by demanding total obedience. The power of the kenotic God lies in giving space for others, dying to the self that others might live. This strange reversal—losing one's life to save it—is also the sensibility that is needed if our planet is to survive and prosper. Giving space is a basic Christian doctrine, but it is also deep in the center of most religions—and it is felt in the hearts of all people, religious or not, who know that love is the discovery of reality, the realization that something beside oneself is real. Other bodies exist and must be fed and cared for. Once that acknowledgement is internalized, there is no going back to the assumption that individuals can pursue their own good apart from the good for others.

The organic model of the world suggests a context for twenty-first-century urban life, a way of thinking, a construction within which to live, that underscores the beauty and intrinsic value of what is left of first nature as well as our inexorable dependence on it. It helps to situate human beings in an appropriate stance toward the world: a stance of gratitude and care, gratitude for the wonder of living on this beautiful planet (as the poet Rilke puts it, "Being here is magnificent"), and care for its fragile, deteriorating creatures and systems. We do not own the earth, we do not even pay rent for it; it is given to us "free" for our lifetime, with the proviso that we treat it with the honor it deserves: appreciating it as a reflection of the divine and loving it as our mother and our neighbor.

Notes

1. This lecture is a revision and update of issues and topics also explored in Sallie McFague, "Where We Live: Urban Ecotheology," in *A New Climate for Theology: God, the World, and Global Warming* (Minneapolis: Fortress Press, 2008); excerpts from chapter 7, 121–139, are used with kind permission of Fortress Press.
2. Thomas Friedman, "The Inflection Is Near?" *New York Times*, March 8, 2009, WK12.
3. See David Owen, *Green Metropolis: Why Living Smaller, Living Closer, and Driving Less Are the Keys to Sustainability* (New York: Riverhead, 2009).
4. Catherine Keller, "Introduction: Grounding Theory: Earth in Religion and Philosophy," in *Ecospirit: Religion, Philosophy, and the Earth,* ed. Laurel Kearns and Catherine Keller (New York: Fordham Univ. Press, 2007), 7.
5. Edward Soja notes the history of these terms: "Hegelian and Marxian notions of 'Second Nature,' a socialized world created from a pristine 'first' Nature but increasingly separate and distinct, subject to its own laws and development," in "Seeing Nature Spatially," paper presented at the University of Chicago Divinity School conference, "Without Nature: A New Condition for Theology," October 26–28, 2006, 2–3. A more down-to-earth definition of "second nature" can be seen in the title of Michael Pollan's book *Second Nature: A Gardener's Education* (New York: Dell, 1991). Here it is not the city as a whole that is second nature, but city gardens in contrast to nature untouched by human hands.
6. Soja, "Seeing Nature Spatially," 13.
7. T. J. Gorringe, *A Theology of the Built Environment: Justice, Empowerment, Redemption* (Cambridge: Cambridge University Press, 2002), 23.
8. As quoted in Soja, "Seeing Nature Spatially," 13.
9. Michael Pollan, *The Omnivore's Dilemma: A Natural History of Four Meals* (New York: Penguin, 2006), 17.
10. Ibid., 9.
11. Ibid., 10.
12. Ibid., 411.
13. Jeffrey Cook, "Environmentally Benign Architecture," *Global Warming and the Built Environment,* ed. Robert Samuels and Deo K. Prasad (London: Spon, 1994), 143.
14. UNEP, Millennium Ecosystem Assessment, Statement from the Board: "Living Beyond Our Means: Natural Assets and Human Well-being," 16.
15. Ibid., 17.
16. Ibid., 5.
17. David Harvey, *Spaces of Hope* (Berkeley: University of California Press, 2000), 220.
18. Peter H. Raven, "The Sustainability of the Earth: Our Common Responsibility," paper presented at University of Chicago Divinity School conference, "Without Nature," 25.
19. See Seppo Kjellberg, *Urban Ecotheology* (Utrecht, the Netherlands: International Books, 2000), 46.

20. Carolyn Merchant, *The Death of Nature: Women, Ecology, and the Scientific Revolution* (San Francisco: Harper and Row, 1983).
21. Harvey, *Spaces of Hope*, 97.
22. For further elaboration, see Sallie McFague, *The Body of God: An Ecological Theology* (Minneapolis: Fortress Press, 1993).
23. John Woolman, *The Journal of John Woolman and Plea for the Poor* (New York: Corinth Books, 1961), 46.
24. J. P. Little, "Simone Weil's Concept of Decreation," *Simone Weil's Philosophy of Culture: Readings Toward a Divine Humanity*, ed. Richard H. Bell (Cambridge: Cambridge University Press, 1993), 25–51.
25. Ibid., 41.
26. Dorothy Day, *The Long Loneliness* (New York: Harper and Row, 1952), 317–318.
27. Stephen Hume, "We Are Seeing the Urban Future and It Is Slums—Slums on a Frightening Scale," *The Vancouver Sun*, June 21, 2006, A17. See the UN report, "State of the World Cities: Globalization and Urban Culture, 2004–05."
28. Report of the Third Session of the World Urban Forum, June 19–23, 2006, Vancouver, BC, 3.
29. "Planetary Boundaries Breached," as quoted in *Vancouver Sun*, September 24, 2009, B5.

An Ethic of the Ordinary
Response to McFague's *Cities, Climate Change, and Christianity*

David C. Lamberth

Taking cities as an orienting problematic for theological reflection, as Professor McFague does in this paper, is both novel and important. The novelty lies in theology's tendency to look away from our proximate environment. But the importance lies in our current circumstances. This is so not only because in 2009 the world crossed a threshold of having half its total population living in cities but also because, as McFague herself says, ecologists agree that whatever their limitations, cities are more efficient habitations for human beings from the energy standpoint. They also have much greater possibilities for increased efficiency in living and flourishing than other environments.

Professor McFague offers a rich and complicated analysis in her paper, noting not only the centrality of cities and the problematic within them of obfuscating their and our dependence on what she calls "first nature." In addition she connects the contributing causes of the climate crisis and the recent economic crisis by noting a common ethical underpinning in an ethic of insatiable desire, development, and growth.

Her theological reflections are equally comprehensive in scope, bringing into relief Christianity's tendency to focus on time and history over space and place, and the modern predominance of the mechanical model of understanding the individual. Notably, she suggests the complicityof this long tradition of reflection and self-understanding in the emergence of current and modern problematics, particularly within typical Western cities.

McFague's prescriptions are also significant, modeling the need to give space, to check desire and self-interest, and crucially emphasizing the need to move from a mechanical model of selves to an organic understanding of the interrelatedness of individuals within cities and the environments and broader nature that subtend and make those lives possible. Turning to the Christian tradition for a key ethical model, Professor McFague takes the lives of extraordinary individuals as a touchstone, highlighting the importance not only of the limit-attending aspects of the prophetic, but also more centrally the place- and space-giving model of kenotic or self-emptying practice, typified by Christ but underlined in figures such as Simone Weil. McFague suggests that an incarnational sense, focused on the embodied and organic nature of us and the world but ethically oriented to the other, might help us navigate the current crisis.

I find all of this extremely productive and appealing and I appreciate very much the deep reflection presented here. However, I do have a few concerns. First, given the scope of the current crisis, can this analysis and ethic be sufficient to orient us toward the changes required in the very near future as we try to reverse our travel beyond the sustainable carbon threshold toward the dreaded tipping point of no return? Bill McKibben's essay in this book mentions that we have reached 390 parts per million of CO_2 in the atmosphere, while best estimates are that 350 is the sustainable level. Second, I am unsure that we do well to identify closely the recent economic crisis with the far more serious and enduring issue of climate change. While both no doubt have to do with the desire for more in the developed West, climate change is driven less by unbridled greed and more fundamentally by the fact that we have structured society's developments around carbon-based energy sources. The many advances that modernity has brought—such as medical developments and improvements in the basic quality of life and extensions of life expectancy—are not themselves features of greed, much less morally problematic, except in terms of their wildly inequitable distribution. Were we to have aligned technology with a fundamentally renewable energy source at some point within the last 250 years, the climate problem would not be what it is. We certainly need to consume less, but the basic needs of 6.7 billion people can only be sustained for the long term by turning away from carbon-oriented technology, not merely limiting our desires severely. This is the underlying imperative, more fundamental than the urban-rural split or the particular way our cities are built, although both of these are subordinately important.

Third, although I find both interesting and inspiring the prophetic kenotic ethics of Jesus and Weil and Woolman, I am concerned that such ethics of extraordinariness may not be able to move us where we need to go. Max Weber, in his analysis of the charisma of the figure of the prophet, notes the extraordinary quality of individuals who occupy these roles, and I daresay the same is true of these inspiring biographies that McFague cites. But is widespread, much less comprehensive, imitation of these actually possible? History teaches us that it may well not be.

I suspect that what is needed is a different ethic of the ordinary, one not too far out of reach of the relatively normal person. Weil's approach to eating, for example, cannot and should not be generalized. Adequate nourishment is key to anyone who does physical labor, such as those who grow our food and make our cities and our lives work, not to mention the malnourished. The exceptional and even elite model here strikes me as impractical, however inspiring it is. And this may well be a problem in general with "*imitatio*" models as an approach, oriented as they so often are toward charismatic and exceptional individuals.

McFague is certainly right that we need a serious reorientation toward one another, toward our shared and communal interest, and between humans and the earth. For other reasons we need to be more egalitarian and more democratic. I am inclined to think that however interesting kenosis is, the more fundamental Christian contribution to ethics, which has comparative analogues in Buddhism and Islam and other religious traditions, is the fundamental religious orientation to the community, and the idea that the individual's true interest is only found in the context of our collective interest. Christianity signifies this initially through the outpouring of the spirit onto the community in Acts (though these days Pentecost is not often read in that fashion). Modern thinkers such as Josiah Royce and Martin Luther King, Jr., continue this communal religious orientation through their talk of building the beloved community in the here and now rather than the there and then. This ethical theme is more fundamentally oriented toward the collective shift, which Professor McFague argues so strongly that we need. More importantly, it is something to which individuals could see their own everyday actions contributing. As much as we like to think that change comes through radical realignments of thought, many ordinary individuals changing their practices incrementally are as crucial, if not more so, from the vantage point of history and psychology.

With a problem so large and immediate as that of current global climate (and the challenge of making livable and equitable cities within that), we can only pray that such religious reevaluations in as many places as possible can combine with other social and political efforts to push us toward an ethical tipping point in terms of our own habits, the ordinary practices that follow on them, and governmental policies. Life itself on the planet does *not* depend on this—I recall quite well Stephen J. Gould in his Ingersoll Lecture pointing out that the height of biodiversity from the standpoint of evolution was 530 million years ago—the time of the flourishing of creatures found now in the Burgess shale near Vancouver. Life will go on even in a carbon cycle that does not support us, so the issue instead is whether we can maintain a place for us, as well as other life. Professor McFague's lecture, above all, reminds us that that remains to be seen.

Human Flourishing Depends on What We Do Now

Bill McKibben

Ecologies of Human Flourishing, the title of this book, sounds so hopeful in the midst of our fears and worries about the planet and the future. How are we going to flourish? That question relates to another which people ask me all the time: What can we do now?

Limits and Their Logic

"What can we do" is a question that we have hardly considered for the last three or four hundred years because the answer to it has been implicit: we can do anything we want and as much of it as we want.

And so we have. We have gone to the moon. We have built enormous homes and enormous vehicles to drive us between them. Many of us fly around the planet again and again for the flimsiest of reasons, appearing hither and yon as if by magic. There has been very little that we could not do in this period because we have had access to cheap fossil fuel, the single most important feature of what we call modern life. Every action of that life is intimately tied to the combustion of fossil fuel twenty-four hours a day, seven days a week—at least in the wealthy parts of the world.

That is now ending. It will end for two reasons. First, we are starting to run short of this magic substance, fossil fuel, which we have been using for several centuries. We do not know how short we are going to run, or exactly how that shortage will manifest itself, but the people who talk about peak oil are correct. We can see the beginning of the end of access to endless cheap supplies of fossil fuel. Second, and more importantly, we are running up against physical limits in the ability of the atmosphere to cope with the byproducts of burning fossil fuel. The gradual accumulation of carbon dioxide (CO_2) in the atmosphere has begun the process of heating the planet with dramatic effect.

Atmospheric chemists and physicists have worried about this possibility for quite a while. Among the first to do so was Arrhenius, the great Swedish scientist, more than a century ago.[1] However, it was only in the 1980s that computer models became sufficiently powerful to offer some sense of what might happen. And only in the last decade have we begun to see with our own eyes the unmistakable effects of that change: for example, the sea ice across the Arctic melting so spectacularly in the summer of 2007[2] and the rapid acidifying of the planet's oceans,[3] something that had not been considered as a possibility until fairly recently. I could elaborate further on climate change and what we are seeing. Suffice it to say that we now have a rough sense of limits.

In 2008 the eminent climatologist Jim Hansen at NASA and a team of fellow scientists published a paper saying that 350 parts per million (ppm) was as much CO_2 as the earth's atmosphere could bear.[4] In the abstract of that journal article, they wrote:

> If humanity wishes to preserve a planet similar to that on which civilization developed and to which life on Earth is adapted, paleoclimate evidence and ongoing climate change suggest that CO_2 will need to be reduced from its current 385 ppm to at most 350 ppm, but likely less than that.[5]

This is strong language for a scientific journal. As of 2010 we are at 390 ppm, and that number is rising steadily. As the abstract (and the article) makes clear, this figure of 350 ppm represents no exact cutoff; it indicates that we have already passed the place where we should be. Whatever the proper limit for CO_2 in the earth's atmosphere, it is behind us now. We need to work at its reduction with much greater urgency.

If we are going to talk about human flourishing, it will need to be within those limits. Given those limits a new logic prevails: our sense of what we need and want out of the world will change. The constant emphasis on endless growth and expansion of economies will seem less desirable and less important. A different set of goals and aspirations will become more powerful. Security and stability will seem more attractive as contrasted to growth, especially in those areas of the world that have already reached the level of affluence necessary to support dignified lives. It will be harder in those areas of the world that have not yet reached that level. In general, the glory associated with the concept of growth will begin to tarnish. Maturity will become our credo instead of growth.

Another logical and even more important development will flow more or less naturally from the beginning of the end of the fossil fuel age. The constant logic of cheap fossil fuel has been to centralize. The logic of what comes next will be to disperse and localize.

To give you a sense of what I am talking about, our food system is now essentially a centralized, industrialized operation. On relatively limited acreage, very few people produce the food we eat. They do so with the help of large quantities of fossil fuel—for the production of fertilizer, for the machines to till the fields, for the trucks to transport food once it is harvested. The food you eat is marinated in crude oil before it reaches your lips. This system will not last. You can already begin to see it breaking down and sense what might come next. Local farmers' markets have been the fastest growing part of the American food economy for the last decade, with the number of markets doubling and doubling again.[6] In some places now local farmers have gone from being a small part of the local food system to becoming a major force in it. For example, in Madison, Wisconsin, on a Saturday the farmers' market often draws 20,000 people.[7] That has been enough to start to rewrite the agricultural economics of that part of the state and change the value and use of land. We are beginning to see that change around the country. The last five-year U.S. agricultural census showed an increase in the number of farms for the first time in about 125 years, most of them small farms growing food for people to eat locally.[8]

Think about energy itself. Under the old model, it made sense to have large centralized power plants. Five or six hundred coal-burning plants scattered around the United States provide about half of its electricity.[9] Fossil fuels are concentrated and rich in BTUs and therefore easy to transport. If we begin to move away from fossil fuels and rely instead on renewable energy, a different set of logical paradigms operates. Sun and wind are widely dispersed; they are available everywhere, but not in large and concentrated quantity. I have solar panels on my rooftop. On a sunny day my house acts as a utility, firing electrons down the grid. You can imagine an energy grid that looks more like a series of farmers' markets than the current centralized model. Of course there will be concentrated pockets of good wind and powerful sun that will be exploited more intensely. In essence, a multimodal energy system emerges—what the engineers call "distributed generation"—as the replacement for today's centralized systems.

Another important commodity on this planet, capital, is also beginning to be distributed, at least in a few cases. Certainly economic events in 2008 and 2009 have helped us understand why we might want to distribute it. The astonishing failure of one big money center, bank, or investment house after another in the last few years, which came very close to bringing the world's economy down with them, reminds us that things that are too big to fail should not be allowed to become that big. We do not need to figure out how to keep them from failing; rather we need to keep them from getting so big that their failure causes massive problems. If you look around the United States, most small banks managed—at least before the ripple effects of the recession caused by the big banks' failures began to work their way down. Local bankers knew to whom they were lending, and what that meant. As a result they have had many fewer problems.[10]

In the world that comes after the fossil fuel era, there will be decreased vulnerability of systems in general because of that decentralization. Take for instance that solar panel on my roof. Theoretically some terrorist could take an interest in it and climb up on my roof and take a hammer to it. But if he did, the only person who would suffer would be me. There would be no deadly solar particles floating out into the atmosphere. The same thing happens in an economy built on small banks. Unwise decisions at First Vermont Middlebury Bank might be enough to cause it and its shareholders problems, but they are not enough to cause problems throughout the entire financial system, as the failure of Lehman and AIG and others clearly did. One experiment in new ways of distributing capital is an alternative complementary currency in Western Massachusetts called BerkShares (after the local mountain range, the Berkshires). There are about three million BerkShares in circulation. You can get them in regular FDIC chartered banks. You can cash your paycheck in U.S. dollars and/or in BerkShares. It helps the money recirculate in the local economy.[11] In the future we will see more of these experiments.[12]

Limits May Help Us Flourish

As this shift to greater localization and less centralization occurs, a new possibility for human flourishing appears, one that has not been available for the last hundred years or so. Fossil fuel has made us lonelier than we were before. It has enabled us to live in ways not possible without access to cheap fossil fuel. With them have come a number of traps into which we have fallen. If you think about the American economy since the end of World War II, by far the biggest part of its vigor has been spent on the project of building bigger homes farther apart from each other, a trend reaching its zenith right before the collapse of the housing market in 2008. Not only did those houses require immense amounts of energy to build, heat, light, and cool (not to mention to drive to and from them), that pattern also had profound effects on who we were. The average American in the 1970s entertained friends and neighbors twice as much as the average American today. American families were three times more likely to eat together in the 1970s than in the late 1990s.[13]

There are many explanations for these differences—television clearly has played a role. However, the mathematical process of isolating ourselves in ever-larger buildings, ever farther apart from each other, has played an enormous role in these deep changes. The number of Americans who say they are very satisfied with their lives peaked sometime in the mid-1950s and has been slowly declining ever since, even as our standard of living has almost tripled.[14] Something is not allowing us to flourish. Life expectancy in the United States now is no longer at the top of the charts.[15] If there are underdeveloped nations in the world there may also be overdeveloped or overconsuming nations in the world, nations of which the United States may be a good example.

Implicit in this new world are some changes that may help us. Here is one statistic that I particularly like. Farmers' markets are the fastest growing part of our food economy and most people think it is because of issues of food quality. But the real reason I think they are growing so quickly is because people are hungry for the contact that farmers' markets provide. Several years ago a pair of sociologists followed shoppers: first around a supermarket, then around a farmers' market. You have been to the supermarket; you know how it works. You walk in, you fall into a light fluorescent trance, you visit the stations of the cross around the supermarket, you emerge somehow with the same basket of goods you had the week before. When these sociologists followed shoppers around the farmers' market, they found something different: shoppers broke up their routines. They stopped to talk, reknitting some of the lost sense of community at the root of our lack of flourishing. Those in the farmers' market averaged ten times more conversations per visit than those in the supermarket.[16] That is to my mind the good news. The bad news is that even as we localize we still have a series of acute global problems that we need to address, climate change being the most important of all.

All of us in the developed world (and Americans in particular) need to learn to do with less, at least in terms of energy. Much of the energy we use is wasted. Some of that waste is technological, and some is lifestyle choice. For example, over the last two decades many Americans have been driving semi-military vehicles back and forth to the grocery store. Once we dispense with the waste, then it gets easier to lower our impact. The average Western European has a higher level of life satisfaction, lives longer, and also uses about half as much energy as the average American.[17] That is a result of different patterns of social organization and community structures. Western Europeans put a higher price on oil at the end of World War II. It was the major reason that they did not sprawl out into suburbs like the United States and that they built and maintained a good rail system. Those indicate some possibilities. Halving energy use will not solve global warming, but it is a start, which technology and other changes can supplement.

Organizing to Address Climate Change

I would like to switch from conjecture about the future and tell you some stories from the last few years of building a movement to create some pressure for action on climate change. Above all this is a political task.

We cannot deal with the problem of climate one community at a time. We do not have the time, and the scale is too large. We must figure out globally and nationally how to put a price on carbon and take effective action to cap its production.

I had no experience building a movement and no idea what I was doing in the fall of 2006. I had just come back from a heart-wrenching trip to Bangladesh. I felt I had to do something, but I did not know what to do: I am a writer; I live in the woods. I called up some friends, other writers in Vermont, and suggested that we stage a sit-in on the steps of the Federal Building in Burlington, the largest city in Vermont. I thought we would be arrested, it would get newspaper coverage, and at least we would have called attention to the issue of global warming. My friends agreed, but one of them called the police in Burlington to ask what would happen. The police said that nothing would happen; we could stay on those steps as long as we wanted. And so we called off that plan and instead marched across Vermont. By the time we finished about a thousand people joined us. This may not sound like very many, but a thousand people is as many people as have ever assembled in Vermont, except for hockey games. Burlington, its largest city, has fewer than 40,000 people and Montpelier, the capital has fewer than 8,000.[18] All the state's candidates for federal office came down to meet with us. Both Republicans and Democrats signed an agreement to work to cut carbon emissions by 80 percent by 2050. The next morning the newspaper said that one thousand people may have been the biggest single demonstration about climate change that had yet taken place in the United States.

I thought then that it was no wonder that climate change was not being addressed by the U.S. government. We had and have the brightest minds in the country at work in Washington and elsewhere with all the policies that we need. But there has been little political force behind them and tremendous force behind those whose interest is served by the status quo. I saw that we needed to draw attention to the issues and to build political capital. We started with college students and an effort called "Step It Up." In the course of a few months, with a few college students we coordinated 1,400 rallies across the United States in the spring of 2007. That helped to sign up both Barack Obama and Hillary Clinton to work for 80 percent cuts in carbon emissions, much more than either of them had ever talked about before. But it clearly was not enough. With the news of the Arctic melt in 2007, the need to work on a planetary scale became even more obvious. We decided to see if we could organize on a global scale. We focused on the number 350 (from the Hansen paper's suggestion of 350 parts of CO_2 per million as a limit) for two reasons: first, it set an actual target so that our leaders would have some accountability, and second, because Arabic numerals can cross linguistic boundaries.

Early on, some old hands in organizing from the 1960s advised us to call for a march on Washington, but we were not willing to ask people to drive across America to protest global warming. Also now the Internet allows us to organize actions on a widespread basis and to coordinate them so that they can have more impact. With "Step It Up," we had beta-tested that idea in the United States; we decided to see if we could do it on a global scale. The crew was seven twenty-three-year-olds, and me, a rapidly aging writer. Seven was a good number: there are seven continents; each of them took a continent. The person who had Antarctica also took the Internet. We began organizing in fall 2008, with the goal of taking the number 350 and inserting it into the conversation about global warming as a way to sharpen the science of the discussion. We picked a date, October 24, 2009, to bring attention to the number 350 parts per million all over the world. We had no idea what might happen. We organized and found excellent people across the globe, but nothing had been done like this. Would it work? We gathered in New York for a final push and to wait for the images from around the world: we had asked local organizers to upload images as soon as their rallies and actions took place.

One of the first indicators we had of success came from Addis Ababa in Ethiopia. It offers an illustration of how we organized. Alia, an eighteen-year old woman, called us from Addis Ababa. She and her sister had attended a climate camp that we offered in South Africa. We trained the participants and they returned to Ethiopia and their homes throughout Africa. We did not hear very much from them because the Internet does not work very well in many parts of Africa. But on that day in October, Alia phoned from Ethiopia to let us know there were 15,000 people demonstrating for 350 in the streets of Addis Ababa. We were very pleased, but because the Internet was down Alia was unable to send us photos. Without a picture we knew that the demonstration would not be covered by news organizations such as AP and Reuters. We located the one Western hotel in Addis frequented by development workers where, finally, through the efforts of three different people, one of them sat at the bar, ordered a drink, and was able to make an Internet connection. A few minutes later we had pictures from Addis, and a few minutes after that we had sent them around the world.

This story of organizing around 350 also follows the distributed model of food, energy, and capital I have been promoting in this lecture. Alia and her sister and their friends organized the demonstration in Addis Ababa, not the eight of us sitting in New York. There were 5,200 events in 181 countries. CNN said it was the most widespread day of political action in the planet's history. *Foreign Policy* called it the largest coordinated global rally ever.[19]

After Alia's photos from Addis Ababa were beamed across the Internet, photos began to pour in from all over the world, many of them so moving that it was difficult to look at them without breaking into tears. There are 25,000 pictures in our Flickr account, http://www.flickr.com/photos/350org/. If you have been told at some point in your life that environmentalists are rich white people, go look at these pictures. Most of them show poor brown, black, Asian, and young people because that is the primary composition of the world's population. They are every bit as interested and concerned as the environmentalists we generally see in the United States—in many cases more so because they already are dealing directly with the effects of global warming.

Let me share with you a few of the projects those pictures illustrate:

- A cooperative project in the Middle East, with people in Jordan forming the three, in Palestine the five, and in Israel the zero, making the point that this work crosses borders;
- Three hundred demonstrations across India organized by young people, including on top of the Red Fort in Delhi and in Bhopal, the scene of one of the largest environmental disasters in the world;
- American soldiers in Afghanistan making a 350 from sandbags—they sent a note saying they were parking their Humvee for the weekend;
- Nobody has Saturday off in Bangladesh, but they stopped work long enough to hold actions all over the country, including making

"350" T-shirts since they are one of the world's largest producers of T-shirts;

- Men in dugout canoes on the Congo River near Conchasa, carrying a 350 banner;
- Huge gatherings in Istanbul: they had had the worst flooding in its history six weeks before;
- Public art in many places, including the Space Needle in Seattle, the London Eye, and the Sydney Opera House, while in Maracaibo, Venezuelans gathered in the colors of the Venezuelan flag to form a 350;
- An orphanage in Indonesia where they picked up bottles—their note said that even though no one cared about them, they cared about the earth;
- Lantern dancers in Australia formed a 350, as did cheerleaders in Syracuse, New York;
- Across Africa, birders joined to identify 350 species that weekend;
- In Soweto in South Africa, they performed 350 bungee jumps, stringing the bungee cord between two cooling towers from a defunct coal-fired power plant and decorated the towers with a 350 banner;
- In China 300 big demonstrations with only one broken up by the police: the young people there knew how to walk up to the line and not cross it—for example, the China Youth Climate Network in Inner Mongolia made a human wind turbine in front of the world's largest turbine field;
- An Iraqi demonstration at the old Hanging Gardens of Babylon;
- Hondurans under military curfew still had a big demonstration;
- In the Maldives the president trained his whole cabinet in scuba diving and they held their cabinet meeting underwater to pass a 350 resolution against the backdrop of their dying coral reef;
- Abu Dhabi has begun to realize they will run out of oil some day and so they have built the world's largest solar array, in front of which there was a 350 demonstration;
- Students drawing 350 pictures at a Pentecostal School in Ghana;
- Wheaton College, where Billy Graham went to school, held a trash audit of dormitories and demonstrations of how to reduce their carbon footprint;
- An interfaith march in South Africa, with the head of Muslim South Africa, a prominent representative of an indigenous religion, and the Anglican Archbishop leading the march through Cape Town.

These last three projects illustrate one of the most encouraging trends that we saw: the deep involvement of religious communities throughout the world. Fifteen years ago or so, there was no general religious environmental movement, but now there is and it is coming of age. Thousands of churches rang their bells 350 times. Buddhists around the world meditated for 350 minutes. Others prostrated themselves 350 times.

Earlier I mentioned the stereotype of environmentalists. There are also stereotypes about those who would not join in such a movement, in particular stereotypes about those in Muslim nations, but we have files and files of pictures of women in burkas who helped to organize events. They care about their children's future, just like everybody else. My favorite leader is Mohamed Nasheed of the Muslim nation of the Maldives. Not only are they performing symbolic actions such as those underwater cabinet meetings, they have also pledged that by 2020 their country will be entirely carbon neutral. They are a poor country, but they are building windmills and tide power plants and other green technologies as fast as they can. They are trying through example to morally persuade the rest of the world to act similarly, because if it does not, their island nation will disappear.

We did not use celebrities. Numbers and scientists were the closest thing we had to celebrities. We took over the big Jumbotrons in Times Square for the day. Instead of the usual advertisements, there were pictures of people from around the world joining one another to save the planet. Almost all the world's papers headlined 350 actions on their front page. The local power that imagined this variety of actions continues to build in these places.

The Political Reality: Bringing Reality to Politics

The response was wonderful, but it has not yet carried the day. At the United Nations Climate Change Conference in Copenhagen six weeks later in December 2009, 117 of the world's nations endorsed the target of 350, many more than we had hoped. However, they were the wrong 117, the poor and vulnerable countries. The rich, powerful, and deeply fossil-fuel addicted countries were not ready to commit to anything at the conference, and they remain in that posture at the moment, the United States among them.

Thus we continue organizing. With the Great Power Race, we are inviting campuses throughout China, India, and the United States to join in a friendly competition to see who can create the most interesting renewable energy projects and sustainability projects on their campuses. In China and in India the world looks very different than it does in the United States. They are excited about the future, and about the prospect of leaping ahead. If some of that human energy and hope can be directed toward clean, green technology, then the planet has a fighting chance.

There will be no silver bullet or bullets, but it may be that there will be enough silver buckshot, a panoply of efforts, lots of bets and avenues to try. That will include a variety of international efforts. Leaders in both China and India clearly understand that there is opportunity; they will try new green technology and methods. For example, Jairam Ramesh, the young environmental minister in India, is positioning his country to take advantage of green movements. It is unclear, however, whether those efforts will be a sideshow to the main event of building more coal-fired power plants, or whether these efforts represent a transformative switch. The Great Power Race and other actions engage young people and their imaginations toward resolving this crisis in the hope of pushing the balance toward transformation.

Another heartening trend internationally is the spread of ideas laterally around the developing world. This year in China several big cities inaugurated bus rapid transit systems, subways on the street. They borrowed this idea not from the United States or even from Europe; they borrowed it from Curitaba in Brazil.[20] In the end, ideas may move from the poor countries to the rich ones. We need mission work going from those places to this country to help us understand the possibilities for the future.

In 2010, rather than a global political rally, we sponsored a global work party all over the world. In thousands of places people installed solar panels, laid out bike paths, dug community gardens—not in the belief that we can solve this problem one project at a time. We cannot: the momentum is too strong, the world too large, but each project helps. More importantly, these projects send a message to our leaders: If we can do this work, then they can do theirs. If we can get up on the roofs of schools and put up solar panels, then people in the Senate can do the work that they are paid to do, the work of writing legislation and passing it. We need to push our leaders much harder.

In the twenty-two years or so of the global warming era, the U.S. Congress has an almost perfect, unblemished, bipartisan record of doing nothing on this question. And the same with many other legislatures. In 2009 the House passed a bill, but it went no further. And it will not, until we build enough pressure to counter that of lobbyists on the other side. We are not going to be able to match the millions of dollars of those lobbyists. Instead, we will have to employ all the creativity, passion, and spirit that people showed in the 350 rallies of October 2009 and 2010 to have a fighting chance.

Unlike other problems to which politicians are accustomed, climate change comes with a time limit. At a certain point in the not-very-distant future it will not matter what the Congress of the United States does or does not do, because the temperature of the planet will be so hot that the permafrost will have begun to melt and methane will be pouring into the atmosphere. Then flexibility and the possibility for action will have disappeared. If we do not deal with climate change soon, we will not deal with it at all. We are used to the political process of bargaining, of compromise. With climate change, the basic negotiation is not between energy companies and environmentalists, the United States and China, Republicans and Democrats. Those are players, but the real negotiation is between human beings on the one hand and physics and chemistry on the other. Physics and chemistry stick to their bottom line. They are unlikely to say: If your economy is in a rough patch we will suspend the laws of nature for a decade until you get it under control. That is one of the reasons that we put this somewhat obscure scientific data point, 350 ppm, at the center of our campaign. There is a reality principle that needs to be respected. We have to measure political process and progress, not against what we have done before, but against whether they rise to what physics demands.

I am not convinced that we are going to succeed; we may have waited altogether too long to get started. But I am convinced that the same model that I talked about in terms of food and energy is the one that we need in terms of politics—people acting in many places and then connecting those actions together and making something bigger than the sum of their parts to create real political pressure. There has to be worldwide political pressure to shift the perceptions and the dynamic. Otherwise there is little likelihood of the major changes we need happening.

The ground for human flourishing is a stable planet. If we cannot do this political work, then all that I have talked about—local food, energy, and capital—will not happen. You can be the best organic gardener in the world, but you need rain. Unless we can do this work very quickly, much else will be impossible. But if we do it right, then much else becomes possible in the process, including the building of meaningful bonds around the world between people of very different circumstances and backgrounds—of different beliefs, classes, and races, to name a few.

It was unbelievably moving to be sitting in New York watching these images of all types of people from all around the world stream across the computer for thirty-six hours. For once, all those clichés about how people are all the same under the skin seemed true.

Notes

1. Arrhenius's 1896 article is one of the first publications on the threat of global warming. See Svante Arrhenius, "On the Influence of Carbonic Acid in the Air upon the Temperature of the Ground," *Philosophical Magazine and Journal of Science* (Fifth Series) 41 (April 1896): 237–276.
2. See Julienne Stroeve, Marika M. Holland, Walt Meier, Ted Scambos, and Mark Serreze, "Arctic Sea Ice Decline: Faster than Forecast," *Geophysical Research Letters* 34 (2007): L09501, doi:10.1029/2007GL029703, in which they suggest that the Arctic may be free of ice by September 2050, if not earlier.
3. For an overview, see Science Policy Section, The Royal Society Policy Report 12/05: "Ocean Acidification Due to Increasing Atmospheric Carbon Dioxide" (London: Royal Society, 2005), for this fact in particular, 11. Available in print and online at the Royal Society website, http://www.royalsoc.ac.uk.
4. James Hansen, Makiko Sato, Pushkher Kharecha, et al., "Target Atmospheric CO_2: Where Should Humanity Aim At?" *The Open Atmospheric Journal* 2 (2008): 217–231.
5. Ibid., 217.
6. Since 1994, when the U.S. Department of Agriculture (USDA) began publishing its national directory of farmers' markets, the number of farmers' markets has grown from 1,755 to 6,132. These numbers are from a chart on growth of farmers' markets prepared by the USDA's Agricultural Marketing Service, updated August 8, 2010, available on the website http://www.ams.usda.gov under "Farmers Markets and Local Food Marketing."
7. R. W. Apple, Jr., "A Peach . . . No, a Honey of a Farmer's Market," *New York Times*, September 29, 2004, F1. Number also confirmed via personal communication with Larry Johnson, manager of the Dane County Farmer's Market, September 2010.
8. Between 2002 and 2007, the number of farms in the U.S. grew from 2,128,982 to 2,204,792 according to the USDA 2007 Census. The average number of acres per farm dropped during that period. See Table 1, Historical Highlights of the USDA 2007 Census Report, available at http://www.agcensus.usda.gov.
9. See "Electric Power Annual with Data for 2008," Washington, D.C.: U.S. Energy Information Administration, released January 21, 2010, http://www.eia.doe.gov/cneaf/electricity/epa/epa_sum.html. See Table 5-1 for number of coal-burning electric power plants. See Figure ES 1, U.S. Electric Power Industry Net Generation, for percentage of U.S. electricity provided by coal (48.2 percent in 2008).
10. For example, a 2009 survey conducted with the help of the Independent Community Bankers of America (ICBA) reported a surge in new banking customers for community banks and an increase of deposits, although there was also impact from the recession on loan portfolios. See Aite Group, *Impact of the Financial Crisis on U.S. Community Banks* (Boston: Aite Group, March 10, 2009, LLC Report #200903092). An executive summary is available on the ICBA website, http://www.icba.org.
11. For more information, see Bill McKibben, "How New England Can Change the World: A Bang for Your Buck in the Berkshires," *Yankee Magazine*, 75th Anniversary Special Issue, September/October 2010, 92.

12. Editors' note: Another of these experiments has been the rise of microfinance, of which the most famous example is probably the Grameen Bank. For more on the Grameen Bank and its founder Muhammad Yunus, see Muhammad Yunus (with Alan Jolis), *Banker to the Poor* (Karachi and New York: Oxford University Press, 2001).
13. See Robert D. Putnam, *Bowling Alone* (New York: Simon and Schuster, 2000), 98–99, 100–101 for these two numbers, in particular. Chapter Six in general focuses on the decline of informal social connections.
14. See Richard Layard, *Happiness: Lessons from a New Science* (New York: Penguin Press, 2005), 29–30, with the happiness scale based on data from Gallup and the income based on data from the U.S. Department of Commerce.
15. *The World Factbook* of the U.S. Central Intelligence Agency gives an estimate of U.S. life expectancy at birth in 2010 as 78.11 years, placing it forty-ninth on its list; this chart is available online at https://www.cia.gov/library/publications/the-world-factbook/rankorder/2102rank.html.
16. Brian Halweil, *Eat Here: Reclaiming Homegrown Pleasures in a Global Supermarket* (New York: Norton, 2004), 10 and 188, note 17, refer both to unpublished studies by Robert Sommer and to the study published in 1981: Robert Sommer et al., "The Behavioral Ecology of Supermarkets and Farmers' Markets," *Journal of Environmental Psychology* 1 (March 1981): 13–19.
17. From *The World Factbook* chart previously cited, life expectancy at birth for the European Union is estimated at 78.82 years for 2010, giving it a ranking of forty-first (with several European countries ranked higher than that, including France, Sweden, Switzerland, Italy, Monaco, Liechtenstein, Spain, Norway, the Netherlands, and Germany). For energy use, see the U.S. Energy Information Administration's website (previously cited), Independent Statistics and Analysis, International Energy Statistics, and search by total primary energy consumption per capita. The site gives the U.S. per capita energy consumption for 2008 as 327 million BTU and Europe's as 143 million BTU.
18. Vermont city population as of 2008, taken from the U.S. Census Bureau webpage, Vermont QuickLinks: http://quickfacts.census.gov/qfd/states/50000lk.html, under "Population Estimates."
19. In its profile of Bill McKibben as number 78 in the Global 100. See *Foreign Policy Special Report:* "The FP Top 100 Global Thinkers," *Foreign Policy,* November 30, 2009, available at http://foreignpolicy.com.
20. For a description of the Curitaba system, see Bill McKibben, *Deep Economy: The Wealth of Communities and the Durable Future* (New York: Times Books/Holt, 2007), 153.

Hope and the Climate Scientist
Response to McKibben's *Human Flourishing Depends on What We Do Now*

Daniel P. Schrag

It is difficult to be a climate scientist these days, watching the decline in both the environment and in American public belief in climate change as an issue. Less than 40 percent of the American public thinks there is solid evidence for global warming due to human activity, according to a 2009 Pew Survey Report.[1] The same report found that only 35 percent believe it is a very serious problem. Those numbers have dropped significantly from 2008, when 44 percent gave that response.[2] The difference can be traced to intensive lobbying—with millions and millions of dollars spent in 2008 both on lobbying on global warming and on advertising.[3] The corporate opposition to this issue is very well organized and powerful; most Americans do not realize how effective it is.

For example, the Intergovernmental Panel on Climate Change (IPCC) has been severely criticized recently. The IPCC is conservative by charter: they do not say anything unless it has been agreed on by hundreds of scientists and is in the published literature. A recent IPCC report said that glaciers in the Himalayas would be gone by 2035.[4] The truth is we do not know how long it will be before those glaciers melt. What the report should have said was that they might be gone by 2035, but there is now an international investigation of the IPCC and its credibility because of this one sentence out of 2,000 pages. That is one indicator of the degree of organization of those opposed to action on climate change. This one sentence has been used to distract attention from the overall problem of climate change and the specific problem of the melting of glaciers.

All over the world, the melting of glaciers threatens agriculture. For example, agriculture in California depends on rivers fed by snow melt from the mountains of the Sierra Nevada. Snow falls in the winter and melts throughout the summer and is a source of water for farming. By the end of the century, maybe sooner, those rivers will run dry during the summer, the peak of the growing season. There is not enough space to build reservoirs to make up for the natural capacity of the mountains. This is true throughout the western United States. The concern in the Himalayas and in the Tibetan plateau is similar; the rivers there sustain agriculture for three billion people.

Bill McKibben gives us hope by reminding us of the involvement of people around the world and the example of the ability to motivate them. He also appeals to human values, values we share. That appeal touches me powerfully, particularly because climate scientists are uncomfortable talking about values. We like to talk about observations. In thinking about values and the value of human flourishing, of simplicity, I am enticed by McKibben's question: could we do less, could we actually live better with less? Part of me aspires to that. Yet, when I think about the world energy systems to which we have become accustomed and which have greatly improved the quality of our life in many ways, part of me worries that less is not enough.

Let me explain what I mean. Many of us, especially those interested in the environment, have a romantic attraction to simplicity, to a life lived simply and in harmony with nature. I suspect we share an uncomfortable intuition that many of the world's problems, including climate change, come from modern technology and the overconsumption of natural resources. If only we could convince everyone to live simpler lives and consume less, perhaps our relationship with nature would move back into balance. At the same time, we should not be naïve about how much we depend on technology. I do not mean only our computers and smart phones, although I am as addicted to those gadgets as anyone. I am thinking more about areas such as health technology. Most people would prefer to live here in Cambridge, Massachusetts, today than to live as an average person in many developing countries because of the differences in infant mortality, adult mortality, and many other health care differences that we take for granted.

In terms of climate change, it turns out that living with less is really not enough. Even if the average person in the world emitted as much greenhouse gases as the average Chinese person—four times less than the average American—we would still have a terrible climate problem. Solving climate change means getting to zero emissions—and that means new technology and lots of it. Convincing people to live with less could reduce energy demand and perhaps slow the rate of greenhouse gas emissions in the future, but I fear that the widespread adoption of zero-carbon energy technology may come quicker in a rapidly growing economy, with people wanting more and better technology. Of course, we still have a lot of work to make sure that the low-carbon technology is indeed cheaper and better. But therein lies the challenge.

I am worried about the tough choices that lie ahead. How will a social movement for acting on climate change deal with those tough choices? Many of the social movements that I have read about and studied over the years have had relatively simple and relatively few ideas or principles to apply. Again with climate change we face some very hard choices and complex tradeoffs. For example Bill McKibben talked about the virtues of distributed energy; in particular he mentioned the solar panels on his home. Solar panels and photovoltaics are expensive; he is fortunate to be able to afford them. Most people in this country would find the cost prohibitive, at least right now. And there are other hard choices. For example, the density at which you can extract energy from the wind is about one watt per square meter, when you average out the spacing of windmills. By way of comparison, the amount of energy you can extract from a square meter of land in a coal mine in Wyoming is approximately a thousand times that of a wind farm, assuming you extract coal over a hundred years. A coal mine is ugly, but it is very dense and confined to a comparatively small area. I am not trying to minimize the horrible effects of burning coal. However, when we talk about wind power as an alternative, we are not talking about building a few windmills, but thousands upon thousands of them. Already Cape Wind in the Nantucket Sound of Massachusetts has garnered strong resistance.[5] Similarly, environmental groups have opposed plans for a solar thermal plant in the Mojave Desert because of concerns over endangered species.[6]

Over the last forty years, environmental groups around the world, but especially in the United States, have defined themselves and built their memberships on the basis of opposition to development. Part of this is grounded in the idea of wilderness and its conservation: essentially environmental groups have lobbied against new building. They have espoused the spirit of what McKibben evoked: let us make do with less.

In contrast, fixing climate change will be a bigger construction project than any other in the last fifty years. It will make building the interstate highway system seem trivial. It is about replacing the infrastructure that has driven the industrial revolution around the world over the last 150 years. And it will cost money: more than 1 percent of Gross Domestic Product (GDP) per year is my guess (and probably closer to 2 percent), but it could be more. For the United States that would be at least two hundred billion dollars a year.[7] We are probably willing to spend twenty billion a year; we are not ready to spend two hundred billion. If the United States spent two hundred billion for thirty years and all the other countries in the world spent proportional amounts, we could fix the problem. It is about building; it is about construction; it is about steel in the ground and concrete and massive building projects—and every single one of them will be opposed by some people who do not want that project located at that particular site. I doubt that Bill McKibben wants windmills throughout the Adirondacks, and neither do I.[8] Wilderness is important, too. If we do not want windmills in the Adirondacks or in Nantucket Sound, where do we agree to put them? These are choices with which we have not yet grappled.

We must have a social movement to support the changes that are needed to address global warming, but it will have to be a social movement unlike any other. Those within it will have to step forward and make difficult decisions, supporting tradeoffs. It is difficult to unify people when they disagree about which tradeoffs should be made. How will we navigate these choices?

It is a tough time to be a climate scientist and tough to be anyone who knows and cares about what is happening to our planet. None of us know whether we will be able to organize and alter human systems soon enough to matter. Despite my concerns and caveats, Bill McKibben's words and actions give us hope for that possibility.

Notes

1. See Pew Survey Report, "Fewer Americans See Solid Evidence of Global Warming," Washington, DC: Pew Research Center for the People & the Press, October 22, 2009, 1; available online at http://people-press.org/report/556/global-warming.
2. Ibid., 2. Similarly, the percentage of those believing that there is evidence of global warming dropped from 71 percent in 2008 to 57 percent in 2009. On the somewhat positive side, the latest survey report shows little decrease in those percentages for 2010, with 59 percent believing in global warming and 34 percent believing global warming is due to human activity. See Pew Survey Report, "Little Change in Opinions about Global Warming," October 27, 2010, 2, http://people-press.org/report/669/.
3. See for example, *The Washington Times,* "Lobbyists See Profit in 'Going Green,'" April 22, 2009, which quotes the Center for Responsive Politics (CRP) as saying that the number of lobbyists working on energy and environmental issues before the U.S. Congress increased to 7,811, an increase of 6 percent over 2007, while related lobbying fees grew to $389 million, an increase of 43 percent over 2007. The same article also discusses an $18 million advertising campaign by the American Coalition for Clean Coal Electricity (ACCE). See also the Center for Public Integrity's website section on the global climate change lobby for a general overview (http://www.publicintegrity.org/investigations/global_climate_change_lobby/). As of the end of 2009, they estimated the number of climate change lobbyists in the United States at close to 3,000, a 400-percent increase since 2003, with 80 percent of the lobbyists working to slow down responses to climate change (Marianne Lavelle and M.B. Pell, "The Climate Lobby from Soup to Nuts," December 27, 2009). See also the CRP's website and its statistics on lobbying by industry sector (http://www.opensecrets.org/lobby/top.php?indexType=i), gathered from the Senate Office for Public Records; for 2010, CRP estimates the energy industry spent $435 million on lobbying (CRP, "Energy and Natural Resouces: Sector Profile, 2010").
4. For the paragraph in the report, see Rex Victor Cruz, Hideo Harasawa, Murai Lal, Shaohong Wu, et al., "Chapter Ten: Asia," *Climate Change 2007: Impacts, Adaptation and Vulnerability, Contribution of Working Group II to the Fourth Assessment Report of the Intergovernmental Panel on Climate Change,* ed. Martin Parry, Osvaldo Canziani, Jean Palutikof, Paul van der Linden, and Clair Hanson (Cambridge: Cambridge University Press, 2007), 493. For an example of the reaction, see Neil MacFarquahar, "Overhaul of U.N. Climate Panel Is Urged," *New York Times,* August 31, 2010, Section A: Foreign Desk, 6.
5. See the website of Save Our Sound, http://www.saveoursound.org. As the main organization of opposition, they list recent news articles from their perspective.
6. See, for example, Felicity Barringer, "Environmentalists in a Clash of Goals," *New York Times,* March 24, 2009, Section A: National, 17, and Todd Woody, "Desert Vistas vs. Solar Power," *New York Times,* December 22, 2009, Section B: Business, 1.
7. The 2010 GDP of the United States is almost 15 trillion dollars; see Bureau of Economic Analysis, U.S. Department of Commerce, Press Release: "Gross Do-

mestic Product: Fourth Quarter and Annual 2010, Advance Estimate," January 28, 2011, 4.

8. Editors' note: In the discussion afterwards, McKibben cited his support for a local windmill project in the Adirondacks because of his fear of the dangers of global warming. He wrote an op-ed in the *New York Times* in support of it. See Bill McKibben, "Tilting at Windmills," *New York Times,* February 16, 2005, Section A: Editorial Desk, 21.

Flourishing and Its Enemies: The Ideology of Self-Interest as Self-Fulfilling

Barry Schwartz

> So far as I am aware, we [i.e., Western society] are the only society that thinks of itself as having arisen from savagery, identified with a ruthless nature. Everyone else believes they are descended from gods. . . . Judging from social behavior, this contrast may be a fair statement of the differences between ourselves and the rest of the world. We make both a folklore and a science of our brutish origins, sometimes with precious little to distinguish between them.
>
> —Marshall Sahlins[1]

Suppose you are directing a daycare center and encounter a problem. The center closes at 6 PM, but a significant number of parents habitually pick up their kids ten to fifteen minutes late. This is justifiably annoying to your staff, which regards lateness as a sign of disrespect. You tack up reminders of the daycare center's hours. You send notices home. But the problem persists. Your warnings become more strident and moralistic, to no avail. You seem to be stuck. Parents know that you cannot close up and leave toddlers outside and alone.

In desperation, you impose a modest fine for lateness, less than the cost of a parking ticket. Will this solve the problem? Uri Gneezy and Aldo Rustichini found out through an experiment at Israeli daycare centers. Prior to the fines, parents were coming late about 25 percent of the time. When the daycare centers introduced fines, the percentage of latecomers *rose*! As the fine imposition continued, lateness continued to increase, almost doubling by the sixteenth week. The daycare centers then discontinued the fines. Lateness increased even more.[2]

Why did the fines have this paradoxical effect? The title of Gneezy and Rustichini's paper tells us: "A fine is a price." We know that a fine is not really a price. A price is what you pay for a service or a good. It is an element of an exchange between willing participants. A fine, in contrast, is a punishment for a transgression. A $25 parking ticket is not the price for parking; it is the penalty for parking illegally. However, the idea of a fine as punishment is not chiseled in stone; there is nothing to stop people from interpreting a fine as a price. If it costs $30 to park in a downtown garage, you may calculate that it is cheaper to park illegally on the street. You are not doing the "wrong" thing; you are doing the economical thing. To get you to stop, we would need to make the fine (price) for parking illegally higher than the price for parking in a garage.

The same phenomenon seems to have occurred in the daycare centers. Originally, parents knew it was wrong to come late. Some parents still came late because other factors outweighed moral considerations. However, they certainly knew that coming late constituted a violation. When the fines were imposed, however, the moral dimension of the parents' behavior disappeared. It became a straightforward economic calculation, and lateness increased. "They're giving me permission to be late. Is it worth $25? Is that a good price to pay to let me stay in the office a few minutes longer? Sure is!" Or perhaps the moral dimension of the behavior did not disappear, but parents judged that the fines constituted enough compensation that lateness was no longer immoral, which is just another way of saying that a fine became a price.

As the daycare center director, you might be tempted to stand on a table and yell at the top of your lungs: "No, no, no! A fine is not a price. A fine does not give you permission to be late! It merely underlines how serious a transgression lateness is." But you would be wasting your breath. The fine gives parents permission to reframe their behavior as an exchange of a fee (the "fine") for a service (fifteen minutes of extra care). Moral considerations are beside the point. Once lost, this moral dimension seems hard to recover. In Gneezy and Rustichini's study, high lateness rates prevailed even four weeks after the fines had been removed. The introduction of fines seemed to permanently alter parents' framing of the situation from a moral transaction to an economic one. When the fines were lifted, lateness simply became a better deal.[3]

Prior to the imposition of fines, parents had one good reason for showing up on time: they felt morally obligated to do so, for coming late would violate a social contract. When the daycare centers introduced fines, a second reason was introduced: not only was lateness a moral violation, but it cost money. Ostensibly, two reasons are better than one, so lateness should have gone down. Instead, it went up. Though it seems perfectly reasonable to assume that motives add—that two reasons are better than one, three are better than two, and so on, this study shows that at least sometimes, motives compete. Adding fines does not give parents a second reason to be on time, but rather it cancels their first reason. In fact, the fines would have had to be draconian to match the moral sanctions that they were replacing.[4]

Competing Motivations

Bruno Frey and Felix Oberholzer-Gee also discovered motivational competition when they assessed the views of Swiss citizens regarding nuclear waste–dump siting. At the time, Switzerland was preparing to have a national referendum about the location of future nuclear waste dumps. Citizens generally held strong views about the issue and were well informed. When asked whether they would be willing to have a waste dump in their community, 51 percent of respondents said yes—despite the fact that people generally thought such a dump was potentially dangerous and would lower their property's value. Nearly 40 percent of respondents believed that the risk of a contaminating accident was significant, and nearly 80 percent of respondents believed that such an accident would result in long-term ill effects for many local residents. However, dumps had to go somewhere, and like it or not, people had obligations as citizens.[5]

Frey and Oberholzer-Gee also asked people a slightly different question. The researchers asked whether people would be willing to have the dumps in their communities in return for an annual payment equivalent to six weeks' worth of an average Swiss salary. Respondents now had two reasons to say yes: obligations as citizens *and* financial incentives. Yet in response to this question, only 25 percent of respondents agreed. Adding the financial incentive cut acceptance in half![6]

Again, it seems self-evident that if people have one good reason to do something and you give them a second reason, they will be more likely to do it. Yet the respondents with two reasons to accept a nuclear waste site were less likely to say yes than those with one. Frey and Oberholzer-Gee explained this result by arguing that financial motives can "crowd out" moral ones. The respondents not offered cash incentives had to decide whether their responsibilities as citizens outweighed their distaste for having nuclear waste dumped in their backyards. Some thought yes, and others, no. But that was the only question they had to answer.[7]

The situation was more complex for the respondents who were offered cash incentives. They had to answer another question before deciding whether to accept the nuclear waste dump in their community: "Should I approach this dilemma as a Swiss citizen or as a self-interested individual? Citizens have responsibilities, but they're offering me money. Maybe the cash is an implicit instruction to answer the question based on the calculation of self-interest." Taking the lead of the questioners, citizens framed the waste-siting issue as an economic issue. With their self-interested hats squarely on their heads, most citizens concluded that six weeks' pay was not enough. In fact, the respondents who rejected the first offer were then offered 50 percent more money, and every respondent but one rejected the second offer as well. The offer of money undermined the moral force of the situation.[8]

These two studies illustrate the central theme of this essay. It is tempting to assume, as economists typically do, that motivation is exogenous to the situation at hand. If you want to influence people to do something, you have to *discover* what motivates them and then structure a situation so that those motives can be satisfied. Related to this deep assumption is the additional one that money—the "universal pander"—is a good proxy for the idiosyncratic motives that individuals possess, because money can be exchanged for almost anything else. In contrast, these two studies suggest that rather than being exogenous to situations, motives can be *created* by situations. Israeli parents do not view their lateness as a market transaction *until* they are fined. Swiss citizens do not view their willingness to accept a waste dump as a market transaction *until* they are offered compensation. Offers of payment or threats of fines do not tap into a motivational structure so much as they *establish* a motivational structure.

This essay also makes a second, related argument. So long as society endorses the legitimacy of different motives for different actions in different situations, Israeli parents or Swiss citizens might ask themselves which motives *ought* to govern their actions in the particular situation they face. However, when one motive gets society's official seal of approval to dominate all others, people may stop appreciating that there are multiple types of motivations from which to choose. Modern Western society's enthusiastic embrace of the view that self-interest simply *is* what motivates human behavior may have led us to create social structures that cater to self-interest. As a result, we have shaped a society in which the assumption that self-interest is dominant is often true. We have not so much discovered the power of self-interest as we have created the power of self-interest. With a debt to Karl Marx, I call such processes "ideology." I believe this ideology, masquerading as truth, poses the most serious obstacle to preserving the planet and the possibility of human flourishing. It is a waste of breath to appeal to people's moral and even spiritual obligations to be stewards of the earth if people no longer comprehend the language of moral and spiritual obligation.[9]

Idea Technology and Ideology

We live in a culture and an age in which the influence of scientific technology is obvious and overwhelming. No one who uses a computer, a smart phone, or an MP3 player needs to be reminded of the power of technology. Nor do people having PET, CAT, and MRI scans, fetuses monitored, genes spliced, or organs transplanted. Adjusting to ever-advancing technology is a brute fact of contemporary life. Some of us do it grudgingly, and some of us do it enthusiastically, but everyone does it.

When we think about the modern impact of science, most of us think about the technology of computers and medical diagnostics—what might be called "thing technology." However, science produces another type of technology that has a similarly large impact on us but is harder to notice. We might call it "idea technology." In addition to creating things, science creates concepts, ways of understanding the world and our place in it, that have an enormous effect on how we think and act. If we understand birth defects as acts of God, we pray. If we understand them as acts of chance, we grit our teeth and roll the dice. If we understand them as the product of prenatal abuse and neglect, we take better care of pregnant women.

If we define technology broadly as the use of human intelligence to create objects or processes that change the conditions of daily life, then it seems clear that ideas are no less products of technology than computers. However, two factors distinguish idea technology from thing technology. First, ideas are intangible and thus cannot be sensed directly. Therefore, they can suffuse the culture and profoundly affect people before being noticed. Second, idea technology, unlike thing technology, can profoundly affect people even if the ideas are *false*. It does so through the process described earlier. Thus I call idea technology based on untrue ideas "ideology." Computers, microwaves, nuclear power plants, and other thing technologies generally do not affect people's lives unless they work. Companies cannot sell useless technological objects—at least not for long. In contrast, untrue ideas can affect how people act as long as people believe them.[10]

Skinnerian Psychology

The potent role of ideology can be illustrated with an example, a critical interpretation of the work of B. F. Skinner that I developed with two colleagues several years ago that relates to the examples that opened this essay.[11] Skinner's central claim was that virtually all animal and human behavior is controlled by its rewarding or punishing consequences. Skinner illustrated this claim with research on pigeons and rats: if a rat receives food pellets consistently after pressing a lever, it will press the lever more often; if the rat receives a painful electric shock after pressing the lever, it stops. For Skinner, the behavior of the lever-pressing rat explains virtually all the behavior of all organisms. To understand behavior, it is necessary and sufficient to identify rewarding and punishing consequences.

Most of Skinner's critics over the years challenged him for being too reductive and for denying the importance, or even the existence, of concepts such as mind, freedom, and autonomy. These critics contended that Skinner's account was not so much false as incomplete and inadequate with regard to human behavior; if one looked with any care at human behavior, one would find numerous phenomena that did not fit the Skinnerian worldview. My colleagues and I took a different approach. We suggested that a casual glance at the nature of life in modern industrial society provided ample justification for the Skinnerian worldview; that is, we agreed with Skinner that virtually all behavior in modern industrial society is controlled by rewards. If one looks at the behavior of industrial workers in a modern workplace, it would be difficult to deny that rats pressing levers for food have a great deal in common with human beings pressing slacks in a clothing factory. Unlike Skinner, however, we argued that this does not reflect basic, universal facts about human nature, but rather reflects the conditions of human labor ushered in by industrial capitalism. We suggested that with the development of industrial capitalism, work was restructured so that it came to look just like rat lever-pressing. The last stages of this restructuring, influenced by the "scientific management" movement of F. W. Taylor, deliberately eliminated all influences on the rate and quality of human labor other than the wage—the reward.[12] Each worker's tasks became so tedious and trivial that he or she simply had no other reasons to work hard. The manager could then exercise complete control over workers by manipulating wage rates and schedules. Skinner developed his theory in a world in which people spent much of their time behaving as he said they would.

Following from our argument, human behavior looks more or less like the behavior of rats pressing levers depending on the structures of the human workplace and other social institutions like schools and mental hospitals. The more these institutions were structured in keeping with Skinner's theory, the truer that theory would look—no, the truer that theory would *be*. Thus, Skinner's theory was ideology—a false piece of idea technology that came to be more and more true as social institutions were shaped in its image.

The examples with which this essay began can be seen as illustrations of this process of ideology in action. If you fine or reward people, their behavior looks as if it is completely governed by fines and rewards. Matters of responsibility—of what is right—will disappear. In a world dominated by incentive manipulations, an open-minded social scientist might conclude, as Skinner did, that the only way to get people to do something is to make it worth their while. This would be correct as a matter of history, but incorrect as a matter of what it means to be human. The examples above are buttressed by a large literature on what is now called the "overjustification effect," which demonstrates how the introduction of rewards for tasks that are normally undertaken without them can change both people's motives to engage in the tasks and the manner in which the tasks are performed.[13]

To summarize, we argued that Skinner's view of human behavior was substantially plausible in the social and economic context in which it arose, though it would not have been plausible in other contexts. Moreover, and more importantly, as the theory was embraced and applied by introducing Skinnerian techniques broadly throughout society, it became more and more plausible. Thus, someone growing up in a post-Skinnerian world in which rewards were routinely manipulated by parents, teachers, clergy, physicians, and law enforcement agents would surely believe that the control of human behavior by such rewards was universal and inevitable. Such a person would be right about the universality, but not about the inevitability.

My colleagues and I were not arguing that simply believing Skinner's worldview was sufficient to make it true. Rather, we argued that believing Skinner's worldview would lead to practices that shaped social institutions in a way that made it true. It is false as a general account of human nature, but as it is embraced and used to reshape one social institution after another, dramatic changes in behavior follow. As a result of this dynamic, an initially false idea—an ideology—becomes increasingly true.

Ideology can become self-fulfilling by several different routes.[14] The one that I believe has the most profound effects is when institutional structures are changed in a way that is consistent with the ideology. The industrialist believes that workers are only motivated to work by wages and then constructs an assembly line that reduces work to such meaningless bits that there is no reason to work aside from the wages. The politician believes that self-interest motivates all behavior, that people are entitled to keep the spoils of their labors, and that people deserve what they get and get what they deserve. Said politician helps enact policies that erode or destroy the social safety net. Unsurprisingly, people start acting exclusively as self-interested individuals: "If it's up to me to put a roof over our heads, put food on the table, and make sure there's money to pay the doctor and the kids' college tuition bills, then I better make sure I take care of myself." Because social structures affect multitudes rather than individuals, we should be most vigilant about the effects of ideology on social structures.

Spheres of Social Life

For the dominance of self-interest to be ideology, there must be alternatives. What are they, and where do they operate? Alan Fiske offered a general framework for answering these questions when he proposed that all societies are governed by four fundamental forms of social relations: communal sharing, authority ranking, equality matching, and market pricing. All four types of relations exist in virtually every society, but societies differ in which of these relations is dominant and in which areas of life are governed by which types of relations. Communal sharing ("what's mine is yours"), authority ranking ("you do what I tell you"), and equality matching ("thanks for mowing my lawn last week when I was away; I'll take care of yours today") are dominant in many societies, whereas market pricing ("what will you pay me to do it") plays a negligible role. But in modern Western societies, market pricing dominates.[15]

Market pricing governs relations in the market, where it is understood that people are interested not in equity, but in gain, and where the principal lubricant of exchange is not trust, authority, or reciprocity, but contracts. Market pricing is facilitated by (perhaps even made possible by) a medium of exchange like money. Money allows people to engage in transactions anonymously and by long distance. As Fiske puts it, transactions need leave no traces, because they can be paid for there and then. There is no need to keep score and no need to expect explicit and direct reciprocation. The market system enables more indirect reciprocation. Buyers do not require the people who sell to them to turn around and buy from them, as long as there are some (potentially anonymous) buyers for every seller. Money also makes it possible to exchange things that would otherwise seem incomparable, or incommensurable. So long as each thing can have a price, then the value of seemingly incomparable things can be compared by means of their prices. This permits a fluidity and freedom in social relations that the other types of social rules do not. With market pricing and a medium of exchange, value is price, everything has a price, and everything can be exchanged for everything else.[16]

Fiske observes that when people deploy one form of social relation when a different one is appropriate, it is often regarded not just as a mistake—a social faux pas—but as a transgression. This moralization has directionality. It is socially acceptable, though perhaps foolish, to rely on communal sharing in a market setting ("Here's a basket of tomatoes; take what you need and pay what you can afford"). But it is taboo to err in the other direction (for example, by giving your mother a tip after a delicious Thanksgiving dinner). Put another way, Fiske's idea that different aspects of life are governed by different norms for appropriate social relations implies that there are barriers that cannot be crossed, goods that are incommensurable, tradeoffs that cannot be made. He refers to such tradeoffs as "taboo tradeoffs."[17] You simply cannot put a price on your mother's Thanksgiving dinner. But the market-pricing model assumes that all goods *are* commensurable. Even goods that are not officially market goods nonetheless have what economists call "shadow prices." Thus, as economist Gary Becker famously argued in his economic analysis of marriage, people stay in unsatisfactory marriages only when the "costs" of breaking up and finding a new partner exceed the "costs" of staying in the marriage.[18] Thus, for example, one can reply to the assertion, "You can't put a price on life" by saying, "Yes you can; we do it all the time." For example, we do not all buy the safest cars. We buy cars that are "safe enough," given what we can afford. If we faced the facts and acknowledge such tradeoffs, would it become easier over time to put a price on life, so that our reverence for life would diminish and it would become just another good?[19] In other words, a little self-deception about the sanctity of life may keep its value higher than it would be otherwise. The same, obviously, can be said when it comes to preserving and sustaining the environment. Thinking of preserving the environment as a sacred responsibility may make environment-consumption tradeoffs much less appealing than simply acknowledging that the tradeoffs are inevitable. This is an argument against, for example, "cap-and-trade" approaches to reducing greenhouse gasses, even when such approaches may, in the short run, reduce greenhouse gases.

In most Western societies, market pricing is the dominant form of social relation—not just in the market, but in public life more generally. The other forms of social relations are mostly relegated to friends and family. But the domains of applicability of these types of relations are dynamic. Over time, as practices evolve, what is taboo becomes acceptable and what is acceptable becomes normal. Though in earlier eras it would have been unthinkable for children to expect payment for doing household chores, this practice has become commonplace. Although it is a topic of current controversy whether students should win cash prizes for school attendance and exam performance, in a somewhat desperate attempt to improve school performance, such practices are now getting serious attention. Twenty years from now, these practices may well have become routine.

One way of understanding the Israeli daycare center and the Swiss nuclear waste referendum examples is that the introduction of fines or the offer of compensation relocated these activities from a different social sphere into the market sphere. Both Israeli and Swiss societies are dominated by market pricing. In a market-pricing dominated world, it takes little effort to shift the character of an activity in that direction. Of course there are some domains of life, even in the United States, that are currently protected from market pricing. However, barriers between domains are not cast in concrete; in the future, market pricing may encroach on even more domains of life. Then we would not have to worry that financial motives would crowd out moral ones, because the moral ones would already have disappeared.

A suggestion that such a change in people's understanding of their social relations and responsibilities to one another can occur on a society-wide scale comes from a recent content analysis of Norwegian newspapers. The analysis, which covered a period from 1984 to 2005, found a shift from what the authors call "traditional welfare ideology," long the dominant sociopolitical characteristic of Norwegian society, to what they call "global capitalist ideology." This shift included increased use of market-like analysis, involving competition and commodification, to discuss all aspects of life; increased reference to the values of individualism and self-interest; and a redefinition of the social contract, along market-pricing lines, between individuals and society.[20] Of course, the fact that newspapers write about social relations in a particular way does not mean that people live them in that way, but it is at least plausible that newspaper coverage either captures a shift in how people think about and act toward one another, or facilitates such a shift, or both.

Dale Miller has presented evidence of the pervasiveness of what he calls the "norm of self-interest" in American society. College students assume, incorrectly, that women will have stronger views about abortion issues than men, and that students under the age of twenty-one will have stronger views about the legal drinking age than those over twenty-one, because women and minors have a stake in these issues that men and older students do not. The possibility that one's views could be shaped by conceptions of justice or fairness, rather than self-interest, does not occur to most people. Miller points out that the self-interest norm has been institutionalized in the legal setting. For instance, courts require parties who write "friend of the court" briefs to have "standing," i.e., an interest in the outcome of the case. Having "skin in the game" is apparently the only way to assure that your opinion will not be frivolous.[21]

Drawing Boundaries between Spheres of Social Life

Given that market pricing is just one of several ways of regulating how people relate to one another, and given that the sorting of life domains into categories of social relation is itself a dynamic process, how should we should decide what goes where? Can normative standards be used to determine whether one mode of social relation is better than another? One could accept, for example, my argument that the dominance of market pricing that we observe in modern Western society is the product of historical contingency—that it has been otherwise in the past, could be otherwise in the future, and is otherwise in other cultures. But one could also argue, as some do, that the dominance of market pricing represents social progress. It caters more than any other system to human freedom and autonomy. It enables people to pursue and get what they want out of life with minimal transaction costs. Exchanges among strangers, which market pricing enables, would not be possible in systems governed by communal sharing ("Who is in your community?"), authority ranking ("From whom do you take orders?"), or equality matching ("I don't have time to mow your lawn. Can I give you a gift certificate to The Home Depot instead?"). Thus, on grounds of freedom, autonomy, and efficiency, one could applaud the dominance of market pricing.

Alternatively, one could argue that although freedom, autonomy, and efficiency are good, they are not the only good. Michael Walzer makes such an argument in his book *Spheres of Justice*. Rather than identifying a single abstract and overarching principle to govern the distribution of goods in society for all spheres of life, Walzer suggests that societies need multiple principles. For example, Walzer argues that in politics, the principle of equality should operate. With regard to basic goods and services, the principle of meeting basic needs should operate. With regard to honors and awards, the operative principle should be merit. As with Fiske's scheme, Walzer acknowledges that boundaries can shift and that principles can be contested. Nonetheless, he argues that thinking about allocating various types of goods requires nuance and subtlety, which the dominance of market pricing threatens.[22]

A second approach to assessing the normative status of market pricing is to accept that freedom, autonomy, and efficiency are important, best achieved by a system of market pricing, but that the effectiveness of market pricing itself depends on commitments to certain moral values. Without the widespread adoption of these moral values, markets will stop working properly and therefore stop serving freedom, autonomy, and efficiency. For example, the productivity that comes from competition can be undermined by the ruthless abuse of market power (involving price fixing and special dealing) by a few market participants. Efficiency can be undermined by massive transaction costs that arise if participants cannot rely on basic honesty. Importantly, values like basic decency and honesty arise out of other systems of social relations. Market pricing does not encourage them, and may erode them. Even Adam Smith, the father of free market economics, held this view. *The Wealth of Nations*, his paean to the marvels of the market, followed another book, *The Theory of Moral Sentiments*, in which he suggested that a certain natural sympathy for one's fellow human beings provided needed restraints on what people would do if they were left free to "barter, truck, and exchange one thing for another." Smith's view, largely forgotten by modernity, was that efficient market transactions were parasitic on aspects of character developed in nonmarket social relations.[23] As I have argued elsewhere, Smith was right about the importance of "moral sentiments," but wrong about how "natural" they are. In a market-dominated society, these "moral sentiments" may disappear so that nothing can rein in self-interest.[24] The same can happen if market activities are walled off from the social relations that exist in other spheres of life, where moral attributes like sympathy might be nurtured.

Echoing this last point, Karl Polanyi suggested that the "great transformation" of society ushered in by industrial capitalism was not so much assembly lines and mass production as it was the separation of the economic sphere of life from other spheres. Economics, Polanyi pointed out, began as "home economics," which meant that economic activity was integrated into the rest of life. With the growth of the factory system, economic activity became increasingly autonomous—separate from the rest of life.[25] As market pricing becomes ever more dominant in regulating social relations, this autonomy hardly matters, since the "rules of the game" outside the market are often the same as the rules of the game inside the market.

The dominance of market pricing represents what one might call "economic imperialism." Can markets continue to provide the benefits they are meant to when they dominate other forms of social relations? President Obama apparently thinks not. In a press conference on December 18, 2008, just before he took office, Obama said that the bankers and financiers must "ask, not only is this profitable . . . but is it right?"[26] In a society dominated by market pricing, where do bankers go to ask "is it right" and find an answer? Each time the scope of the market extends itself, it becomes increasingly difficult for people to envision an alternative to its logic. As a result, the self-interest assumptions built into market pricing become "laws of nature" by default. And we see the consequences. We have all been suffering through the consequences of a financial system completely untethered from any concern about "is it right?"

Conclusion

In his book *A Conflict of Visions,* Thomas Sowell distinguishes between what he calls "constrained" and "unconstrained" visions of human nature. The constrained vision, exemplified by Thomas Hobbes, focuses on the selfish, aggressive side of human nature; it argues that we cannot change human nature but must instead impose constraints through an all-powerful state, the Leviathan. The unconstrained vision, best exemplified, perhaps, by Jean-Jacques Rousseau, sees enormous human possibility, and condemns the state for subverting the good in human nature.[27] This essay has argued that both Hobbes and Rousseau are wrong. "Nature" dramatically underspecifies human nature. Within broad limits, we are what society asks and expects us to be. If society asks little, it gets little. Under these circumstances, we must be sure that we have arranged social rules and incentives in a way that induces people to act in ways that serve the common good. If we ask more of people and arrange our social institutions correctly, we will get more. As Geertz has said, human beings are "unfinished animals," and what we can reasonably expect of people depends on how our social institutions "finish" them.[28] Market capitalism is not psychologically inert. It does not simply take people as they are and cater to their goals and desires. It promotes some goals and desires and minimizes others. It encourages values like materialism, individualism, and competition that compete with and crowd out other values that may better serve both societies and individuals.

Why do people support government policies that encourage individual economic entrepreneurship and material consumption and neglect other values that contribute to well-being? Why do people insist, despite massive evidence that above subsistence, additional material wealth makes a trivial contribution to well-being, that the only thing standing between them and perfect happiness is a few more dollars in their pay envelopes?[29] A recent paper by Daniel Kahneman and others suggests an answer to this question. People are committing what has come to be called the "focusing illusion," focusing on one determinant of well-being to the exclusion of all others.[30] But why are they focusing on wealth, rather than meaningful work, stewardship of the earth, or close social relations?

The answer, I fear, was provided by Margaret Thatcher when she was Prime Minister of Great Britain—TINA: There Is No Alternative.[31] Twenty pages every day in the newspaper are regularly devoted to the health of financial institutions, none to that of civic institutions. We have cabinet-level ministers for finance, but not for well-being. We measure per capita GDP, but not per capita GDW (gross domestic well-being). Wherever we turn, the lesson taught is that what we do and should care about is ourselves, and more specifically, about our material selves. We are living in a monoculture.

Rational, self-interested, economic man as a reflection of human nature is a fiction—an ideology. But it is a powerful fiction, and it becomes less and less fictional as it pervades our institutions and crowds out other types of social relations. Because of its self-fulfilling character, we cannot expect this fiction to die of natural causes. To extinguish it, we must hold onto the alternatives. This will not be easy.

Human beings are remarkably good at adapting to the contexts within which they live. A generation raised in the context of close social ties and civic engagement will miss these nurturing aspects of life as market capitalism corrodes them. But this generation will adapt. The next generation, born into an environment that is thoroughly individualistic and materialistic, will simply regard it as natural.

More is going on than mere adaptation. Because capitalism is corrosive, it is likely that nonmaterialistic and nonindividualistic aspects of life will become less good—less easy to realize and less satisfying—as their contact with market institutions continues. People will then feel like they are losing less, and thus be relatively uninterested in challenging capitalism's monolithic stature. I have suggested this possibility in two books.[32] Hochschild's brilliant discussion of the "commercialization of feeling" captures this danger well. As Hochschild points out, as more and more Americans earn their livelihoods in service industries, the product they have to sell is themselves. As they compete with one another for customers, the premium goes to the ones who can be most "sincere" in catering to the desires and interests of customers.[33] But once deep emotions become instrumentalized in this way, people may no longer be able to distinguish the genuine from the fake—both in others and in themselves (as suggested by a quote variously attributed to Daniel Schorr, George Burns, and others: "The secret to success in this business is sincerity. If you can fake that, you've got it made"). With only ersatz emotional connection available from others, why not plow straight ahead into a world of individualist, material consumption? There is not much to be lost. If you make the "goods" that are incompatible with capitalism less good, individualistic materialism becomes the only game in town: There Is No Alternative.

Although people are adaptable, there are some living conditions to which people should not be asked to adapt. The single-mindedness of market capitalism is one of them. Suppose a society were to decide to cut off everyone's right arm. In the short run, this would create great misery, but people would adapt. The manufactured world would be re-engineered to make it possible for people to live their lives successfully with one arm. The next generation, with right arms severed at birth or perhaps tied behind their backs, would never know there was anything missing—that once, more was possible in life. Our task, I believe, is to help prevent market capitalism from cutting off people's right arms. We have to do it before people reach a point where the capitalist way of life seems natural—indeed, seems to be the only possible way of life. We know now that "There Is No Alternative" is false. Our grandchildren might not.

In the *Nicomachean Ethics*, Aristotle famously wrote about *eudaimonia*, or what we might now call "flourishing."[34] For Aristotle, flourishing demanded virtue. It demanded that people do the right thing, and that they do it for the right reason. Material incentives may sometimes get people to do the right thing, but decidedly not for the right reasons. Kenneth Sharpe and I have recently attempted to bring Aristotle's thinking to bear on twenty-first-century life. [35] Contemporary research in psychology strongly suggests that for people to flourish, they need meaningful, engaging work and close social relations. [36] What Sharpe and I argue is that for people to have meaningful, engaging work and close social relations, they need virtue. We argue, in short, that Aristotle was right. If we aspire to human flourishing, we must bring the moral dimensions of life back into focus, and refuse to settle for "efficient," market-driven alternatives.

Notes

1. Marshall D. Sahlins, *The Use and Abuse of Biology: An Anthropological Critique of Sociobiology* (Ann Arbor: University of Michigan Press, 1976), 100.
2. Uri Gneezy and Aldo Rustichini, "A Fine Is a Price," *Journal of Legal Studies* 29 (January 2000): 1–17.
3. Ibid.
4. Uri Gneezy and Aldo Rustichini, "Pay Enough or Don't Pay at All," *Quarterly Journal of Economics* 115 (August 2000): 791–810.
5. Bruno Frey and Felix Oberholzer-Gee, "The Cost of Price Incentives: An Empirical Analysis of Motivation Crowding-Out," *American Economic Review* 87 (September 1997): 746–755.
6. Ibid.
7. Ibid.
8. Ibid.
9. For a similar argument in the context of legal decision making, see Barry Schwartz, "Crowding Out Morality: How the Ideology of Self-Interest Can Be Self-Fulfilling," in Jon Hanson, ed., *Psychology, Ideology, and Law* (New York: Oxford University Press, forthcoming).
10. For further elaboration, see Barry Schwartz, "Psychology, Idea Technology, and Ideology," *Psychological Science* 8, no. 1 (January 1, 1997): 21–27.
11. See Barry Schwartz, Richard Schuldenfrei, and Hugh Lacey, "Operant Psychology as Factory Psychology," *Behaviorism* 6 (1978): 229–254. See also Barry Schwartz, *The Battle for Human Nature: Science, Morality and Modern Life* (New York: Norton, 1986); "Some Disutilities of Utility," *Journal of Thought* 23 (1988): 132–147; and "The Creation and Destruction of Value," *American Psychologist* 45 (1990): 7–15.
12. F. W. Taylor, *The Principles of Scientific Management* (1911, New York: Norton, 1967).
13. See Mark R. Lepper, David Greene, and Richard E. Nisbett, "Undermining Children's Intrinsic Interest with Extrinsic Rewards: A Test of the 'Overjustification' Hypothesis," *Journal of Personality and Social Psychology* 28, no. 1 (1973): 129–137; and Mark R. Lepper and David Greene, eds., *The Hidden Costs of Reward: New Perspectives on the Psychology of Human Motivation* (Hillsdale, NJ: Lawrence Erlbaum Associates, 1978). See also Barry Schwartz, "Reinforcement-Induced Behavioral Stereotypy: How Not to Teach People to Discover Rules," *Journal of Experimental Psychology: General* 111, no. 1 (March 1982): 23–59.
14. See Schwartz, "Psychology, Idea Technology, and Ideology."
15. See Alan Fiske, *Structures of Social Life: The Four Elementary Forms of Human Relations* (New York: Free Press, 1991), and "The Four Elementary Forms of Sociality: Framework for a Unified Theory of Social Relations," *Psychological Review* 99, no. 4 (October 1992): 689–723.
16. Fiske, *Structures of Social Life.*
17. Alan Fiske and Philip E. Tetlock, "Taboo Trade-offs: Reactions to Transactions that Transgress the Spheres of Justice," *Political Psychology* 18 (1997): 255–297.

18. Gary Becker, Elisabeth M. Landes, and Robert T. Michael, "An Economic Analysis of Marital Stability," *Journal of Political Economy* 85, no. 6 (1977): 1143–1175.
19. See Schwartz, "Disutilities of Utility."
20. Hilde Nafstad, Rolv Blakar, Erik Carlquist, Joshua Phelps, and Kim Rand-Hendriksen, "Globalization, Neo-liberalism, and Community Psychology," *American Journal of Community Psychology* (February 2009), doi: 10.1007/s10464-008-9216-6.
21. Dale Miller, "The Norm of Self-Interest," *American Psychologist* 54 (1999): 1053–1060.
22. Michael Walzer, *Spheres of Justice: A Defense of Pluralism and Equality* (New York: Basic Books, 1983).
23. Adam Smith, *The Theory of Moral Sentiments* (1753, Oxford, UK: Clarendon Press, 1976) and *The Wealth of Nations* (1776, repr., New York: Modern Library, 1937), also available online as a Pennsylvania State University Electronic Classic, http://www2.hn.psu.edu/faculty/jmanis/adam-smith/Wealth-Nations.pdf, quoted material is from this edition, 18.
24. See Barry Schwartz, *The Costs of Living: How Market Freedom Erodes the Best Things in Life* (New York: W. W. Norton, 1994) and *Battle for Human Nature.*
25. See Karl Polanyi, *The Great Transformation: Economic and Political Origins of Our Time* (New York: Rinehart, 1944).
26. President Barack Obama, press conference, December 18, 2008.
27. Thomas Sowell, *A Conflict of Visions: Ideological Origins of Political Struggles* (New York: William Morrow, 1987).
28. Clifford Geertz, *The Interpretation of Cultures* (New York: Basic Books, 1973), 49.
29. See Ed Diener and Martin E. P. Seligman, "Beyond Money: Toward an Economy of Well-Being," *Psychological Science in the Public Interest* 5, no. 1 (2004): 1–31 for a review of the evidence.
30. Daniel Kahneman, Alan B. Krueger, David Schkade, Norbert Schwarz, and Arthur A. Stone, "Would You Be Happier If You Were Richer? A Focusing Illusion," *Science* 312 (2006): 1908–1910, doi: 10.1126/science.1129688.
31. See, for example, Margaret Thatcher's June 25, 1980, press conference for American correspondents available on the Thatcher Foundation website, http://www.margaretthatcher.org. *There Is No Alternative* also is the title of a recent favorable biography of Thatcher by Claire Berlinski (New York: Basic Books, 2008).
32. See Schwartz, *Battle for Human Nature* and *Costs of Living.*
33. See Arlie Russell Hothschild, *The Managed Heart: Commercialization of Human Feeling* (Berkeley: University of California Press, 1983).
34. Aristotle, *Nicomachean Ethics*, trans. Martin Ostwald (New York: Library of Liberal Arts, 1962).
35. Barry Schwartz and Kenneth Sharpe, *Practical Wisdom: The Right Way to Do the Right Thing* (New York: Riverhead, 2010).
36. See Martin E. P. Seligman, *Authentic Happiness* (New York: Free Press, 2002), and Ed Diener and Robert Biswas-Diener, *Happiness: Unlocking the Mysteries of Psychological Wealth* (New York: Blackwell, 2008).

Notes on Contributors

Volume Editors

Donald K. Swearer is Distinguished Visiting Fellow, Center for the Study of World Religions (CSWR) at Harvard Divinity School (HDS), and Professor Emeritus of Religion, Swarthmore College. He served as director of the CSWR and Distinguished Visiting Professor of Buddhist Studies at HDS for six years. Before coming to HDS in 2004, Professor Swearer taught at Swarthmore College from 1970 to 2004, from 1992 to 2004 as the Charles and Harriet Cox McDowell Professor of Religion. In 2010–11 he conducted research on Buddhist economics and Thailand's sufficiency economy as a Fulbright Senior Research Scholar at Payap University, Chiang Mai, Thailand. His most recent books include *Rethinking the Human* (2010) and *Ecology and the Environment* (2009), both of which he edited; and *The Buddhist World of Southeast Asia,* second revised edition (2009). Other recent books include *Sacred Mountains of Northern Thailand and Their Legends* (2004), *Becoming the Buddha: The Ritual of Image Consecration in Thailand* (2004), and *The Legend of Queen Cama: Bodhiramsi's Camadevivamsa, a Translation and Commentary* (1998). Professor Swearer has published several essays on Buddhism and ecology and is a founding board member of the Forum on Religion and Ecology.

Susan Lloyd McGarry, managing editor, Office of Communications, HDS, shepherded three other CSWR books (*Rethinking the Human,* 2010; *Ecology and the Environment,* 2009; and *Religion and Nationalism,* 2006) through editing and production with Donald K. Swearer, before moving to the Office of Communications where she is one of the editors of *Harvard Divinity Bulletin*, a general-interest magazine on religion. As assistant director of planning and special projects at the CSWR, she managed a faculty grant program, publications, an international visiting scholar program, and an international student internship, as well as acting as liaison to the CSWR Advisory Board. A published poet, her poems have been anthologized in *The Poetry of Peace* (2001) and she was named Bard of the 2004 Boston Irish Festival for her poem, "Memory of Coumenole." Prior to coming to Harvard in 2001, she served as executive director of the Quaker retreat center, Woolman Hill.

Contributors

Steven B. Bloomfield is executive director of Harvard's Weatherhead Center for International Affairs where he has also been associate director and director of its Fellows Program. He has served as a director of scholarships for Latin Americans studying in the United States, taught in the Boston Public Schools, worked in rural development and education as a Peace Corps volunteer in the Ecuadorian Andes, instructed in the elementary grades in a New York City independent school, and served as a member of a settlement-house staff and as a community organizer in Manchester and London. He holds degrees from Harvard's Graduate School of Education and Harvard College.

Lawrence Buell is Powell M. Cabot Professor of American Literature at Harvard University, where he has taught since 1990. His research interests include rethinking U.S. literature in a globalizing world, discourses of literature and environment, and the theory of national fiction. Buell began writing about Thoreau in his first book, *Literary Transcendentalism* (1973), a critical and historical analysis of the literary side of the Transcendentalist movement, a nineteenth-century New England effort to develop an ecology of human flourishing. Issues of place, community, and ideologies of belonging have been central to Buell's books, including *New England Literary Culture* (1986), *The Environmental Imagination: Thoreau, Nature Writing, and the Formation of American Culture* (1995), *Writing for an Endangered World: Literature, Culture, and Environment in the United States and Beyond* (2001), *Emerson* (2003), and *The Future of Environmental Criticism* (2005). *Writing for an Endangered World* won the Popular Culture and American Culture

Associations' Cawelti Prize for the best book of 2001 in the field of American cultural studies. In 2007, the Modern Language Association honored Professor Buell with the Jay Hubbell Award for lifetime contributions to American literary studies.

Diana L. Eck, Professor of Comparative Religion and Indian Studies and Frederic Wertham Professor of Law and Psychiatry in Society at Harvard University, chairs Harvard's Department of Sanskrit and Indian Studies. She serves as a member of the Faculty of Divinity and the Committee on the Study of Religion, as well as Master of Lowell House, an undergraduate residence. Diana Eck's academic work concentrates on India and the United States. Her work on India focuses on popular religion. She also teaches a seminar, "Gandhi: Then and Now." Her books include *Banaras: City of Light* (1982) and *Darsan: Seeing the Divine Image in India* (1981; reissued 1996). Her current book project, "India: A Sacred Geography," forthcoming in 2011, is on networks of pilgrimage in India. Diana Eck's work on the United States focuses on the challenges of religious pluralism in a multireligious society and includes the award-winning *Encountering God: A Spiritual Journey from Bozeman to Banaras* (1993) and *A New Religious America: How a "Christian Country" Has Become the World's Most Religiously Diverse Nation* (2001).

Paul Farmer, M.D., is the Koloktrones University Professor of Global Health and Chair of the Department of Global Health and Social Medicine at Harvard Medical School; Chief of the Division of Global Health Equity at Brigham and Women's Hospital; and cofounder of Partners In Health. He also serves as United Nations Deputy Special Envoy for Haiti, under Special Envoy Bill Clinton. Dr. Farmer and his colleagues have pioneered novel community-based treatment strategies that demonstrate the delivery of high-quality health care in resource-poor settings. He has written extensively on health, human rights, and the consequences of social inequality. His most recent book is *Partner to the Poor: A Paul Farmer Reader* (2010). Other titles include *Pathologies of Power: Health, Human Rights, and the New War on the Poor* (2003); *Infections and Inequalities: The Modern Plagues* (1999); and *The Uses of Haiti* (1994). He is a member of the Institute of Medicine of the National Academy of Sciences and of the American Academy of Arts and Sciences. He holds a doctorate in anthropology from Harvard, as well as an M.D.

Bridget Hanna is a PhD candidate in Medical Anthropology at Harvard University. Her dissertation research is on changes in contemporary Indian medical practice in response to the environmental health crisis. She is the author of several articles, coeditor of *The Bhopal Reader* (2005), and founder of the Bhopal Memory Project. She is also an activist and a documentary filmmaker. Her most recent film *This Much I Know* has been shown at film festivals internationally. More information about her work can be found at http://harvard.scholar.edu/bridgethanna.

Arthur Kleinman, M.D., is one of the world's leading medical anthropologists, and a major figure in cultural psychiatry, global health, and social medicine. He is the Esther and Sidney Rabb Professor, Department of Anthropology, Harvard University; Professor of Medical Anthropology in Social Medicine and Professor of Psychiatry, Harvard Medical School; and Victor and William Fung Director of Harvard University's Asia Center. Since 1968, Kleinman, who is both a psychiatrist and an anthropologist, has conducted research in Chinese society, first in Taiwan, and since 1978 in China, on depression, somatization, epilepsy, schizophrenia and suicide, and other forms of violence. His chief publications are *Patients and Healers in the Context of Culture* (1980); *Social Origins of Distress and Disease: Depression, Neurasthenia, and Pain in Modern China* (1986); *The Illness Narratives (1988); Rethinking Psychiatry* (1988); and the co-edited volumes, *Culture and Depression* (1985) and *Social Suffering* (1997). His most recent book, *What Really Matters* (2006), addresses existential dangers and uncertainties that make moral experience, religion, and ethics so crucial to individuals and society today.

David C. Lamberth, HDS Professor of Philosophy and Theology, joined the HDS faculty in 1997 as an assistant professor, after teaching at Florida State University. His courses in Western theology and the philosophy of religion emphasize modern liberal thought and probe the interconnections between theological and philosophical reflection. His 1999 *William James and the Metaphysics of Experience* exhibits his interest in the revival of pragmatism, and demonstrates the inherent engagement with religion in James's philosophical system, as well as James's pluralism. He is currently preparing two books: "Religion: A Pragmatic Approach," which analyzes both historical and contemporary treatments of religion in the pragmatic tradition, and a volume on William James for the Routledge Philosophers series.

Sallie McFague is currently Distinguished Theologian in Residence, Vancouver School of Theology. She formerly served as Dean of Vanderbilt Divinity School, as well as teaching there for thirty years. She has been a leader in linking ecofeminism and Christian theology, known for her working with new metaphors. Her most recent book is *A New Climate for Theology: God, the World, and Global Warming* (2008). Among her many influential works are *Life Abundant: Rethinking Theology and Economy for a Planet in Peril* (2000); *Super, Natural Christians: How We Should Love Nature* (1997); *The Body of God: An Ecological Theology* (1993); *Models of God: Theology for an Ecological, Nuclear Age* (1987), which received the American Academy of Religion's Award for Excellence; and *Metaphorical Theology: Models of God in Religious Language* (1982).

Bill McKibben, Schumann Distinguished Scholar at Middlebury College, is an American environmentalist and author who frequently writes about global warming, alternative energy, and local economies. He has led one of the largest demonstrations against global warming in American history and continues to organize around this issue. In January 2007 he founded Step It Up 2007 to demand that Congress enact curbs on carbon emissions that would cut global warming pollution 80 percent by 2050. With many of the youth organizers involved in Step It Up, he cofounded and directs 350.org, an international campaign named for the parts per million of CO_2 that scientists have identified as the safe upper limit for our planet. McKibben's influential first book, *The End of Nature* (1989), discussed climate change and led the way for public discussions of global warming. Among his recent books are *Eaarth: Making a Life on a Tough New Planet* (2010), *The Bill McKibben Reader* (2008), and *Deep Economy: The Wealth of Communities and the Durable Future* (2007).

Anne E. Monius, Professor of South Asian Religions at HDS, is a historian of religion specializing in the religious traditions of India. Her research interests lie in examining the practices and products of literary culture to reconstruct the history of religions in South Asia. Her first book, *Imagining a Place for Buddhism: Literary Culture and Religious Community in Tamil-Speaking South India*, examines the two extant Buddhist texts composed in Tamil. Her current research project, "Singing the Lives of Siva's Saints: History, Aesthetics, and Religious

Identity in Tamil-Speaking South India," considers the role of aesthetics and moral vision in the articulation of a distinctly Hindu religious identity in twelfth-century South India. Both works point to a larger research focus on the ways in which aesthetics and ethics define religious identity and community in South Asia, as well as to the creative and productive encounters among competing sectarian religious communities.

Chandra Muzaffar, president of the International Movement for a Just World (JUST), an international organization based in Malaysia, which critiques global injustice and seeks to develop an alternative vision of a just and compassionate civizilation, is also the Noordin Sopiee Professor of Global Studies at the Science University of Malaysia (USM) in Penang. He previously served as the director for the Centre for Civilisational Dialogue, University of Malaya. Professor Muzaffar has published extensively on religion, human rights, Malaysian politics, civilizational dialogue, and international relations. He has authored or edited 26 books and written more than 500 articles in English and Malay in various local and international journals. His most recent books include *Exploring Religion in Our Time* (2011), *Hegemony: Justice, Peace* (2008), and the edited volume *Religion Seeking Justice and Peace* (2010). Among the academic awards Chandra Muzaffar has received is the Harry J. Benda Prize for distinguished scholarship on Southeast Asia from the Association of Asian Studies, North America.

Daniel P. Schrag is Professor of Earth and Planetary Sciences, Harvard University, and director of the Harvard University Center for the Environment. He studies climate and climate change over the broadest range of Earth history. A former MacArthur Fellow, he is widely published in scientific journals. He has examined changes in ocean circulation over the last several decades, with particular attention to El Niño and the tropical Pacific. He has worked on theories for Pleistocene ice-age cycles including a better determination of ocean temperatures during the Last Glacial Maximum, 20,000 years ago. Schrag also helped develop the Snowball Earth hypothesis, proposing that a series of global glaciations occurred between 750 and 580 million years ago that may have led to the evolution of multicellular animals. Currently he is working with economists and engineers on technological approaches to mitigating future climate change.

Barry Schwartz is the Dorwin Cartwright Professor of Social Theory and Social Action in the Psychology Department at Swarthmore College, where he has taught since 1971. He does research on decision making, and has written extensively about the intersection of psychology, economics, and morality. Among other books, he has written *The Battle for Human Nature: Science, Morality, and Modern Life* (1986), *The Costs of Living: How Market Freedom Erodes the Best Things in Life* (1994), *The Paradox of Choice: Why More Is Less* (2004), and most recently, with Kenneth Sharpe, *Practical Wisdom: The Right Way to Do the Right Thing* (2010).

Ronald F. Thiemann holds the Benjamin Bussey Professorship of Theology, the oldest endowed chair in theology at Harvard. An ordained Lutheran and a specialist on the role of religion in public life, Professor Thiemann has been at Harvard since 1986 and served as Dean of the Divinity School from 1986 until 1998. Among his several books are *Religion in Public Life: A Dilemma for Democracy* (1996), *Constructing a Public Theology: The Church in a Pluralistic Culture* (1991), *Revelation and Theology: The Gospel as Narrated Promise* (1985, reissued 2005), and the edited volume, *Who Will Provide? The Changing Role of Religion in American Social Welfare* (2001). He is currently working on a book-length project entitled "Sacramental Realism: Literary Art as Social Critique," which will trace the relations among religion, aesthetics, and politics in both visual and literary realist art.

Archana Venkatesan is Assistant Professor of Comparative Literature and Religious Studies at the University of California, Davis. She is the recipient of numerous grants, including fellowships from the National Endowment for the Humanities, American Institute of Indian Studies, and Fulbright. Her research interests are in the intersection of text and performance in South India as well as in the translation of early and medieval Tamil poetry into English. She is the author of *The Secret Garland: Āṇṭāḷs Tiruppāvai and Nācciyār Tirumoḻi* (Oxford University Press, 2010) and the forthcoming *A Hundred Measures of Time: Nammāḻvār's Tiruviruttam.*

Index

nature (*continued*)
  and Garden of Empowering Liberation, 3–4
  human dependence on, forgotten, 130–131
  order in, and ethics in medieval South India, 43–47, 50–53, 63–64, 69
  "second," 130–134, 138, 143–144
  value of, 3, 134
Nearing, Helen and Scott, 33
neutrality, illusion of, 94
Nhat Hanh, Thich, 5–6, 8–9
Nicomachean Ethics (Aristotle), 191
nirvana, 116
No Impact Man (Beavan), 24–25, 31, 35
norms, 95, 105, 184–186
  American, of self-interest, 185
Norway, 185
nuclear waste, disposal of, 177–178, 185

Obama, President Barack, 159, 188
Oberholzer-Gee, Felix, 177–178
oil
  and environment, 101, 154–155, 158, 162
  and the Middle East, 105–106
organic model of the world, 137–138, 144–145
Out of My Life and Thought (Schweitzer), 26
overjustification effect, 181
Owl (Maotouying) (Huang), 87

Palestine, 106, 161
Pallavas, 40–53, 61, 63–64, 69
  anxiety about drought and famine, 43
  destroyers of forests, 46
  devotional literature of, 40
  ethics and the natural world among, 43–47, 50–53, 69
  and mediating role, 40, 42, 50, 52
  name of means "leaf," 61
  Rājasiṁha (king), 47–53
  and religious diversity, 40, 52
  temples of, 40, 44–45, 48–52, 68
  and water, 42–48, 50–53
Panamalai, Tālapurīśvara Temple, 48–49, 52
Pandian, Anand, *Crooked Stalks*, 69
paradigm. *See* Kuhn, Thomas
Paradox of Choice, The (Schwartz), 7
Partners In Health (PIH), 80, 98–99n9
Pathologies of Power (Farmer), 80, 94
patriotic shopping, 15
Peking Union Medical College, 79, 80
PEPFAR, 81
personal efficacy, 85, 94–95, 97
Pēy, 64–66
Philippians, 138
"Please Call Me by My True Names" (Nhat Hanh), 8
Polanyi, Karl, 187–188
politics, and climate change, 158–159, 163–166, 169, 171–173
Pollan, Michael, 133
polls
  American on global warming, 169
  American on materialism, 14, 16
  Swiss on nuclear waste dumps, 177–178, 185
poor, the
  and economic injustice, 36–37, 108–110
  and old attitudes of deserving vs. undeserving, 21–22
  preferential option for, 80–81, 140
  rights of, 92
  service to, and personal efficacy, 85, 94–95, 97
Poor Richard's Principle (Wuthnow), 18
power, 96
  counterhegemonic, and religion, 126–127
  and religious elites, 113, 120, 121
Poykai, 64–66
practices, spiritual, of resistance, 127
prayer, 33, 114–115
preferential option for the poor, 80–81
principles, first, 86, 92–96
prison, as barometer for society's moral health, 93
Protestantism
  as prophetic stream of Christianity, 138, 142, 150, 151
  and work ethic, 18
  See also Christianity *and under specific denominations*
psychophysiology, of religious inspiration, 83–86
Pūtam, 64–66
Putnam, Robert, *Bowling Alone*, 6